TestSMART®

for
Math Concepts

Grade 4

Help for
Basic Math Skills
State Competency Tests
Achievement Tests

by

Lori Mammen

These popular teacher resources and activity books are available from ECS Learning Systems, Inc. for Grades K–6.

Basic Skills Software for Math	Gr. 1-10	240 Lessons
Basic Skills Software for Reading	Gr. 1-10	405 Lessons
The Bright Blue Thinking Books™	Ages 6–12	3 Titles
Building Language Power	Gr. 4–9	3 Titles
Foundations for Writing	Gr. 2–8	2 Titles
Get Writing!!™	Gr. K–5	6 Titles
Graphic Organizer Collection	Gr. 3–12	1 Title
Home Study Collection™	Gr. 1–6	18 Titles
Inkblots™	Gr. K–6	2 Titles
The Little Red Writing Books™	Ages 6–12	3 Titles
Math Whiz Kids™	Gr. 3–5	4 Titles
Novel Extenders™	Gr. 1–6	7 Titles
On the Chalkboard™	Gr. 1–6	6 Titles
Once Upon a Time™ for Emerging Readers	Gr. K–2	10 Titles
Once Upon a Time™ (Books + Tapes)	Gr. K–2	10 Titles
The Picture Book Companion	Gr. K–3	3 Titles
Quick Thinking™	Gr. K–6	1 Title
Springboards for Reading	Gr. 3–6	2 Titles
Structures for Reading, Writing, Thinking	Gr. 4–9	4 Titles
Test Preparation Guides	Gr. 2–12	41 Titles
Wake Up, Brain!!™	Gr. 1–6	6 Titles
Writing Warm-Ups™	Gr. K–6	2 Titles

To order, or for a complete catalog, write:

ECS Learning Systems, Inc.
P.O. Box 791439
San Antonio, Texas 78279-1439
Web site: **www.educyberstor.com**

Cover: Kirstin Simpson
Book Design: Educational Media Services

ISBN 1–57022–242-8

Contents

Introduction   4

What's inside this book?   4

How to Use This Book   6

Master Skills List   7

Master Skills Correlation Chart   8

Number Concepts   9

Mathematical Relations, Functions, & Algebraic Concepts   33

Geometric Properties/Relationships   49

Measurement Concepts   83

Probability and Statistics   103

Appendix   123
    Answer Key   124
    Answer Sheets   127

Welcome to *TestSMART*®!!

It's just the tool you need
to help students review important mathematics skills and
prepare for standardized mathematics tests!

Introduction

During the past several years, an increasing number of American students have faced some form of state-mandated competency testing in mathematics. While several states use established achievement tests, such as the Iowa Test of Basic Skills (ITBS), to assess students' achievement in mathematics, other states' assessments focus on the skills and knowledge emphasized in their particular mathematics curriculum. Texas, for example, has administered the state-developed Texas Assessment of Academic Skills (TAAS) since 1990. The New York State Testing Program began in 1999 and tests both fourth- and eighth-grade students in mathematics.

Whatever the testing route, one point is very clear: the trend toward more and more competency testing is widespread and intense. By the spring of 1999, 48 states had adopted some type of assessment for students at various grade levels. In some states, these tests are "high-stakes" events that determine whether a student is promoted to the next grade level in school.

The emphasis on competency tests has grown directly from the national push for higher educational standards and accountability. Under increasing pressure from political leaders, business people, and the general public, policy-makers have turned to testing as a primary way to measure and improve student performance. Although experienced educators know that such test results can reveal only a small part of a much broader educational picture, state-mandated competency tests have gained a strong foothold. Teachers must find effective ways to help their students prepare for these tests—and this is where *TestSMART*® plays an important role.

What's inside this book?

Designed to help students review and practice important math and test-taking skills, *TestSMART*® includes reproducible practice exercises in the following areas—

- pretests for each of the five major objectives addressed in the book
- practice exercises that target the specific skills tested within each objective

In addition, each *TestSMART*® book includes—

- a master skills list based on mathematics standards of several states
- complete answer keys

The content of *TestSMART*® is outlined below.

Major Objectives: This book focuses on five major mathematics objectives which represent broad areas of understanding generally common to all grade levels. These objectives focus on developing students' understanding of—

• number concepts
• mathematical relations, functions, and other algebraic concepts
• geometric properties and relationships
• measurement concepts using metric and customary units
• probability and statistics

Specific Skills: A list of specific skills appears below each major objective. These skills represent appropriate grade-level expectations for a given objective.

Pretests: Five pretests are included in this book. Each pretest addresses one of the major objectives and includes test items for all the specific skills included with that objective. Teachers use the pretests to diagnose the students' areas of strength and weakness for a given objective.

Practice Exercises: Practice exercises follow each pretest. Unlike the pretests, each practice exercise addresses a specific skill. Teachers use the practice exercises to target specific areas of weakness revealed in the pretests.

Master Skills List/Correlation Chart: The mathematics skills addressed in *TestSMART*® are based on the mathematics standards and/or test specifications from several different states. No two states have identical wordings for their skills lists, but there are strong similarities from one state's list to another. The Master Skills List for Mathematics (page 7) represents a synthesis of the mathematics skills emphasized in various states. Teachers who use this book will recognize the skills that are stressed, even though the wording of a few objectives may vary slightly from that found in their own state's test specifications.

The Master Skills Correlation Chart (page 8) offers a place to identify the skills common to both *TestSMART*® and a specific state competency test. To show how such a correlation can be done, the author has included a sample correlation which shows the skills addressed in both *TestSMART*® and the skills tested on the Texas Assessment of Knowledge and Skills (TAKS).

Answer Keys: Complete answer keys appear on pages 124–126.

How to Use This Book

Effective Test Preparation: What is the most effective way to prepare students for any mathematics competency test? Experienced educators know that the best test preparation includes three critical components—

- a strong curriculum that includes the content and skills to be tested
- effective and varied instructional methods that allow students to learn content and skills in many different ways
- targeted practice that familiarizes students with the specific content and format of the test they will take

Obviously, a strong curriculum and effective, varied instructional methods provide the foundation for all appropriate test preparation. Contrary to what some might believe, merely "teaching the test" performs a great disservice to students. Students must acquire knowledge, practice skills, and have specific educational experiences which can never be included on tests limited by time and in scope. For this reason, books like *TestSMART*® should **never** become the heart of the curriculum or a replacement for strong instructional methods.

Targeted Practice: *TestSMART*® does, however, address the final element of effective test preparation (targeted test practice) in the following ways—

- *TestSMART*® familiarizes students with the content usually included in competency tests.
- *TestSMART*® familiarizes students with the general format of such tests.

When students become familiar with both the content and the format of a test, they know what to expect on the actual test. This, in turn, improves their chances for success.

Using *TestSMART*®: Used as part of the regular curriculum, *TestSMART*® allows teachers to—

- pretest skills needed for the actual test students will take
- determine students' areas of strength and/or weakness
- provide meaningful test-taking practice for students
- ease students' test anxiety
- communicate test expectations and content to parents

Master Skills List

I. Demonstrate an understanding of number concepts
A. Use place value to read, write, compare, and order whole numbers through the millions place
B. Determine the value of a number written in expanded notation
C. Read, write, compare, and order fractions (like and unlike denominators) and decimals (to thousandths)
D. Name and write mixed numbers as improper fractions and improper fractions as mixed numbers
E. Relate decimals to fractions that name tenths and hundredths (with and without models)

II. Demonstrate an understanding of mathematical relations, functions, and other algebraic concepts
A. Identify and extend whole-number and geometric patterns
B. Identify number sentences that show the inverse relationships between addition/subtraction and multiplication/division (fact families)
C. Identify the relationship between two sets of related data, such as ordered pairs in a table

III. Demonstrate an understanding of geometric properties and relationships
A. Identify right, acute, obtuse, and straight angles
B. Identify models of intersecting, parallel, and perpendicular lines
C. Identify the parts of a circle (center, radius, chord, diameter)
D. Identify and describe shapes and solids in terms of their properties (sides, vertices, edges, faces)
E. Identify congruent shapes
F. Identify lines of symmetry in shapes
G. Locate and name points on a number line using whole numbers, fractions (halves and fourths), and decimals (tenths)

IV. Demonstrate an understanding of measurement concepts, using metric and customary units
A. Estimate, measure, and compare weight using customary and metric units
B. Estimate, measure, and compare capacity using customary and metric units
C. Estimate, measure, and compare length using customary and metric units
D. Carry out simple unit conversions within a system of measurement

V. Demonstrate an understanding of probability and statistics
A. List possible outcomes of a probability experiment (e.g., tossing a coin)
B. Make predictions based ona sampling
C. Determine the mean (average), median, and mode from collected data
D. Intepret bar graphs, tables, and charts

Master Skills Correlation Chart

Use this chart to identify the *TestSMART*® skills included on a specific state competency test. To correlate the *TestSMART*® skills to a specific state's objectives, find and mark those skills common to both. The first column shows a sample correlation based on the Texas Assessment of Knowledge and Skills (TAKS).

Sample
Correlation
↓

	Sample Correlation	
I. Demonstrate an understanding of number concepts		
A. Use place value to read, write, compare, and order whole numbers through the millions place	★	
B. Determine the value of a number written in expanded notation		
C. Read, write, compare, and order fractions (like and unlike denominators) and decimals (to thousandths)	★	
D. Name and write mixed numbers as improper fractions and improper fractions as mixed numbers		
E. Relate decimals to fractions that name tenths and hundredths (with and without models)	★	
II. Demonstrate an understanding of mathematical relations, functions, and other algebraic concepts		
A. Identify and extend whole-number and geometric patterns		
B. Identify number sentences that show the inverse relationships between addition/subtraction and multiplication/division (fact families)	★	
C. Identify the relationship between two sets of related data, such as ordered pairs in a table	★	
III. Demonstrate an understanding of geometric properties and relationships		
A. Identify right, acute, obtuse, and straight angles	★	
B. Identify models of intersecting, parallel, and perpendicular lines	★	
C. Identify the parts of a circle (center, radius, chord, diameter)		
D. Identify and describe shapes and solids in terms of their properties (e.g., sides, vertices)		
E. Identify congruent shapes	★	
F. Identify lines of symmetry in shapes	★	
G. Locate and name points on a number line using whole numbers, fractions (halves and fourths), and decimals (tenths)	★	
IV. Demonstrate an understanding of measurement concepts, using metric and customary units		
A. Estimate, measure, and compare weight using customary and metric units	★	
B. Estimate, measure, and compare capacity using customary and metric units	★	
C. Estimate, measure, and compare length using customary and metric units		
D. Carry out simple unit conversions within a system of measurement		
V. Demonstrate an understanding of probability and statistics		
A. List possible outcomes of a probability experiment (e.g., tossing a coin)	★	
B. Make predictions based on a sampling		
C. Determine the mean (average), median, and mode from collected data		
D. Intepret bar graphs, tables, and charts	★	

Number Concepts

I. Demonstrate an understanding of number concepts

A. Use place value to read, write, compare, and order whole numbers through the millions place

B. Determine the value of a number written in expanded notation

C. Read, write, compare, and order fractions (like and unlike denominators) and decimals (to thousandths)

D. Name and write mixed numbers as improper fractions and improper fractions as mixed numbers

E. Relate decimals to fractions that name tenths and hundredths (with and without models)

Notes

Objective 1: Pretest

I.A *Use place value to read, write, compare, and order whole numbers through the millions place (1-10)*

1. Which group of numbers is in order **from least to greatest**?

   A 2,314 1,461 3,810 1,384

   B 1,461 1,384 2,314 3,810

   C 1,384 1,461 2,314 3,810

   D 3,810 2,314 1,461 1,384

2. Which group of numbers is in order **from greatest to least**?

   A 1,816 1,842 1,801 1,839

   B 1,842 1,839 1,816 1,801

   C 1,839 1,816 1,842 1,801

   D 1,801 1,816 1,839 1,842

3. The chart shows the population in four counties.

County	Population
Bear	14,338
Clark	10,562
Dawson	15,125
Falls	13,625

   Which shows the counties in order **from greatest to least** population?

   A Dawson, Falls, Bear, Clark

   B Bear, Falls, Dawson, Clark

   C Clark, Falls, Bear, Dawson

   D Dawson, Bear, Falls, Clark

4. Which number is greater than 7,104 but less than 7,160?

   A 7,041

   B 7,099

   C 7,154

   D 7,200

5. What is the value of 4 in 12,473?

   A Ones

   B Hundreds

   C Tens

   D Thousands

6. Which number has a 6 in the thousands place?

   A 563

   B 4,612

   C 26,904

   D 30,106

7. Which number has a 9 in the ten thousands place and a 1 in the hundreds place?

   A 94,163

   B 91,030

   C 9,103

   D 94,310

8. The Nile River is 4,145 miles long. How is this number read?

   A Four thousand, one hundred forty-five

   B Forty thousand, one hundred forty-five

   C Forty-one thousand, forty-five

   D Four million, one hundred forty-five

9. A forest covers fifty-eight thousand, four hundred seventy-five acres. How is this number written?

   A 5,8400,075

   B 5,804,475

   C 580,475

   D 58,475

10. Which number is 2,000 **greater** than 12,132?

    A 12,332

    B 32,132

    C 14,132

    D 10,332

I.B **Determine the value of a number written in expanded notation (11-14)**

11. Which one shows 12,071?

    A $1,200 + 70 + 1$

    B $1,000 + 200 + 0 + 1$

    C $120 + 70 + 1$

    D $10,000 + 2,000 + 70 + 1$

12. Find the number that is the same as $2,000 + 400 + 30 + 6$.

    A 24,036

    B 2,436

    C 20,436

    D 240,306

13. Which one shows 6,158?

    A $6000 + 100 + 50 + 8$

    B $610 + 50 + 8$

    C $600 + 150 + 8$

    D $1000 + 600 + 50 + 8$

14. Find the number that is the same as $40,000 + 2000 + 300 + 10 + 4$.

    A 4,214

    B 420,314

    C 42,314

    D 402,314

earning Systems, Inc.

I.C Read, write, compare, and order fractions [like and unlike denominators] and decimals [to thousandths] (15-26)

15. Which figure has $\frac{1}{5}$ shaded?

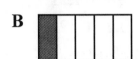

16. Which group shows the fractions in order **from least to greatest**?

A $\frac{1}{2}$ $\frac{1}{3}$ $\frac{1}{4}$ $\frac{1}{5}$

B $\frac{1}{3}$ $\frac{1}{5}$ $\frac{1}{2}$ $\frac{1}{4}$

C $\frac{1}{4}$ $\frac{1}{5}$ $\frac{1}{3}$ $\frac{1}{2}$

D $\frac{1}{5}$ $\frac{1}{4}$ $\frac{1}{3}$ $\frac{1}{2}$

17. Which group shows the fractions in order **from greatest to least**?

A $\frac{1}{6}$ $\frac{1}{3}$ $\frac{1}{4}$ $\frac{1}{8}$

B $\frac{1}{3}$ $\frac{1}{4}$ $\frac{1}{6}$ $\frac{1}{8}$

C $\frac{1}{8}$ $\frac{1}{4}$ $\frac{1}{6}$ $\frac{1}{3}$

D $\frac{1}{8}$ $\frac{1}{6}$ $\frac{1}{4}$ $\frac{1}{3}$

18. Which group shows the fractions in order **from greatest to least**?

A $\frac{2}{5}$ $\frac{2}{3}$ $\frac{1}{4}$ $\frac{1}{2}$

B $\frac{1}{4}$ $\frac{2}{5}$ $\frac{1}{2}$ $\frac{2}{3}$

C $\frac{2}{3}$ $\frac{1}{4}$ $\frac{2}{5}$ $\frac{1}{2}$

D $\frac{2}{3}$ $\frac{1}{2}$ $\frac{2}{5}$ $\frac{1}{4}$

19. The figures are shaded to show equivalent fractions.

Which is equivalent to $\frac{2}{3}$?

A $\frac{2}{6}$ C $\frac{4}{6}$

B $\frac{3}{6}$ D $\frac{3}{4}$

20. Which group shows $\frac{3}{6}$ of the boxes shaded?

21. Which decimal is equal to two tenths?

A 0.02

B 0.2

C 0.12

D 2.1

22. Which group shows the decimals in order **from greatest to least**?

A 0.12   0.21   0.1   0.01

B 0.1   0.12   0.01   0.21

C 0.21   0.12   0.1   0.01

D 0.21   0.1   0.12   0.01

23. Which group shows the decimals in order **from least to greatest**?

A 0.05   0.1   0.15   0.2

B 0.1   0.2   0.05   0.15

C 0.2   0.15   0.1   0.05

D 0.05   0.2   0.15   0.1

24. Which of these is true?

A 0.03 > 0.3

B 0.18 < 0.1

C 0.4 < 0.06

D 0.09 < 0.15

25. How is the decimal 0.12 read?

A twelve tenths

B one and two tenths

C twelve hundredths

D twelve thousandths

26. Which group shows the decimals in order **from greatest to least**?

A 0.12   0.102   0.2   0.125

B 0.125   0.102   0.12   0.2

C 0.2   0.102   0.12   0.125

D 0.2   0.125   0.12   0.102

I.D   Name and write mixed numbers as improper fractions and improper fractions as mixed numbers (27-32)

27. Which improper fraction equals $1\frac{1}{4}$?

A $\frac{2}{4}$

B $\frac{4}{4}$

C $\frac{5}{4}$

D $\frac{6}{4}$

28. Which mixed number equals $\frac{6}{5}$?

A $1\frac{1}{5}$

B $1\frac{1}{6}$

C $2\frac{1}{5}$

D $1\frac{2}{5}$

29. Which mixed number represents the shaded parts of the boxes?

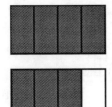

A $1\frac{1}{2}$

B $2\frac{1}{4}$

C $1\frac{2}{3}$

D $1\frac{3}{4}$

30. Which improper fraction represents the shaded parts of the boxes?

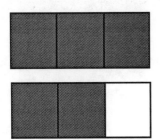

A $\frac{5}{2}$

B $\frac{4}{3}$

C $\frac{5}{3}$

D $\frac{6}{3}$

31. Which mixed number equals $\frac{7}{3}$?

A $1\frac{1}{3}$

B $2\frac{1}{3}$

C $2\frac{1}{2}$

D $3\frac{1}{3}$

32. Which improper fraction equals $3\frac{1}{3}$?

A $\frac{10}{4}$

B $\frac{9}{3}$

C $\frac{10}{3}$

D $\frac{11}{3}$

I.E *Relate decimals to fractions that name tenths and hundredths [with and without models] (33-40)*

33. Which figure has $\frac{1}{10}$ shaded?

A

B

C

D

34. Which decimal equals $\frac{15}{100}$?

   A 0.15

   B 0.015

   C 1.05

   D 0.115

35. Which decimal square is $\frac{1}{2}$ shaded?

A

B

C

D

36. Which fraction is equal to 0.2?

   A $\frac{2}{5}$

   B $\frac{1}{2}$

   C $\frac{2}{10}$

   D $\frac{2}{100}$

37. Which decimal is equal to $\frac{4}{100}$?

   A 0.4

   B 0.14

   C 0.41

   D 0.04

38. Which fraction tells how much of the figure is shaded?

   A $\frac{4}{100}$

   B $\frac{40}{100}$

   C $\frac{14}{100}$

   D $\frac{50}{100}$

39. Which fraction is equal to 0.35?

   A $\frac{3}{5}$

   B $\frac{30}{100}$

   C $\frac{35}{100}$

   D $\frac{35}{10}$

40. Which decimal is equal to $\frac{52}{100}$?

   A 0.52

   B 0.502

   C 0.052

   D 0.512

Practice 1.A1

1. Which group of numbers is in order **from least to greatest**?

   A 9,594   9,700   9,682   9,710

   B 9,700   9,710   9,682   9,594

   C 9,594   9,682   9,700   9,710

   D 9,682   9,710   9,594   9,700

2. Which group of numbers is in order **from greatest to least**?

   A 1,128   1,234   1,138   1,255

   B 1,234   1,255   1,128   1,138

   C 1,255   1,234   1,138   1,128

   D 1,128   1,138   1,234   1,255

3. The chart shows the number of books borrowed from 4 libraries during one year.

Library	Books Borrowed
Gray	24,075
Hayes	23,908
Jones	24,109
Madison	23,861

   Which shows the libraries in order **from greatest to least** number of books borrowed?

   A Jones, Gray, Hayes, Madison

   B Jones, Hayes, Gray, Madison

   C Hayes, Madison, Jones, Gray

   D Gray, Jones, Madison, Hayes

4. Which number is less than 112,603 but greater than 112,547?

   A 112,500

   B 112,550

   C 112,610

   D 112,650

5. What is the value of 6 in 610,943?

   A Hundreds

   B Thousands

   C Ten thousands

   D Hundred thousands

6. Which number has a 7 in the tens place and a 4 in the millions place?

   A 4,591,275

   B 4,169,527

   C 4,007,265

   D 406,275

7. The number 1,246 is read—

   A twelve thousand, twenty-four six

   B one thousand, twenty-four six

   C one thousand, two hundred six

   D one thousand, two hundred forty-six

Practice 1.A2

I.A   Use place value to read, write, compare, and order whole numbers through the millions place

1. Which group of numbers is in order **from least to greatest**?

   A 9,701   7,109   1,970   1,079

   B 1,970   9,701   1,079   7,109

   C 1,079   1,970   7,109   9,701

   D 1,970   1,079   7,109   9,701

2. Which number is 2,000 more than 13,475?

   A 13,495

   B 33,475

   C 13,675

   D 15,475

3. The chart shows the number of students in 4 school districts.

District	Students
Clarke	10,286
James	10,826
Martin	10,682
Johnson	10,268

   Which shows the districts in order **from least to greatest** number of students?

   A Johnson, Clarke, Martin, James

   B Martin, James, Johnson, Clarke

   C Clarke, Martin, James, Johnson

   D James, Clarke, Martin, Johnson

4. Which number is greater than 5,013 but less than 5,124?

   A 5,134

   B 5,120

   C 5,010

   D 5,142

5. What is the value of 8 in 8,162,379?

   A Thousands

   B Ten thousands

   C Hundred thousands

   D Millions

6. Which number has a 3 in the tens place and a 1 in the ten thousands place?

   A 612,348

   B 621,438

   C 836,912

   D 912,436

7. Members of a hiking club hiked sixteen thousand, four hundred ten miles. How is this number written?

   A 1,641

   B 16,041

   C 16,410

   D 160,410

Practice 1.A3

1. Which group of numbers is in order **from least to greatest?**

   A 2,319   2,391   2,193   2,139

   B 2,139   2,193   2,319   2,391

   C 2,193   2,139   2,319   2,391

   D 2,391   2,319   2,193   2,139

2. Which number is 4,000 more than 12,098?

   A 13,875

   B 43,475

   C 12,138

   D 16,098

3. The chart shows the number of books read by 4 classes in one school year.

Teacher	Books Read
Potter	1,080
Garcia	1,801
James	1,108
Thomas	1,008

   Which shows the classes in order **from least to greatest** number of books read?

   A Thomas, Potter, James, Garcia

   B James, Garcia, Potter, Thomas

   C Potter, Garcia, Thomas, James

   D Garcia, James, Potter, Thomas

4. Which number is greater than 12,122 but less than 12,301?

   A 12,098

   B 12,310

   C 12,112

   D 12,294

5. What is the value of 4 in 14,338?

   A Tens

   B Hundreds

   C Thousands

   D Ten thousands

6. Which number has a 2 in the millions place and a 7 in the ten thousands place?

   A 1,207,943

   B 1,276,349

   C 2,176,349

   D 2,716,439

7. Students at Smith Elementary School collected one hundred eighteen thousand, four hundred sixteen pounds of newspaper. How is this number written?

   A 100,018,416

   B 1,180,416

   C 1,018,416

   D 118,416

Practice 1.B1

I.B *Determine the value of a number written in expanded notation*

1. Which one shows 32,801?

    A 3,200 + 800 + 1

    B 3,000 + 200 + 80 + 1

    C 30,000 + 200 + 800 + 1

    D 30,000 + 2,000 + 800 + 1

2. Which one shows 7,615?

    A 700 + 60 + 10 + 5

    B 7,000 + 60 + 10 + 5

    C 7,000 + 600 + 10 + 5

    D 6,000 + 700 + 10 + 5

3. Which one shows 8,430?

    A 8,000 + 400 + 30

    B 4,000 + 800 + 30

    C 800 + 400 + 30

    D 8,000 + 400 + 3

4. Find the number that is the same as 20,000 + 8,000 + 100 + 50 + 6.

    A 2,856

    B 280,156

    C 20,856

    D 28,156

5. Find the number that is the same as 1,000 + 300 + 30 + 9.

    A 13,339

    B 1,339

    C 1,309

    D 130,339

6. Find the number that is the same as 40,000 + 1,000 + 70 + 2.

    A 4,172

    B 40,172

    C 41,072

    D 410,072

7. Which one shows 9,531?

    A 9,000 + 500 + 30 + 1

    B 9,000 + 50 + 30 + 1

    C 900 + 500 + 30 + 1

    D 95,000 + 30 + 1

8. Which one shows 11,207?

    A 1,000 + 200 + 7

    B 10,000 + 1,000 + 200 + 7

    C 10,000 + 1,000 + 200 + 70

    D 10,000 + 200 + 7

Practice 1.B2

I.B   *Determine the value of a number written in expanded notation*

1. Which one shows 6,128?

   A 6,000 + 12 + 8

   B 1,000 + 600 + 20 + 8

   C 6,000 + 100 + 20 + 8

   D 60,000 + 1,000 + 200 + 80

2. Find the number that is the same as 60,000 + 100 + 90 + 3.

   A 61,093

   B 60,193

   C 6,193

   D 60,903

3. Find the number that is the same as 8,000 + 700 + 4.

   A 8,704

   B 80,704

   C 874

   D 87,004

4. Find the number that is the same as 90,000 + 1,000 + 400 + 50 + 7.

   A 910,457

   B 914,057

   C 9,557

   D 91,457

5. Which one shows 5,621?

   A 5,000 + 60 + 21

   B 50,000 + 6,000 + 200 + 10

   C 5,000 + 600 + 20 + 1

   D 6,000 + 500 + 20 + 1

6. Which one shows 14,219?

   A 1,000 + 4,000 + 200 + 10 + 9

   B 10,000 + 4,000 + 200 + 10 + 9

   C 10,000 + 4,000 + 200 + 9

   D 100,000 + 4,000 + 200 + 10 + 9

7. Which one shows 80,374?

   A 80,000 + 300 + 70 + 4

   B 8,000 + 300 + 70 + 4

   C 8,000 + 1,000 + 300 + 70 + 4

   D 80,000 + 3,000 + 700 + 4

8. Find the number that is the same as 10,000 + 6,000 + 500 + 1.

   A 160,501

   B 16,501

   C 1,651

   D 16,051

Practice 1.B3

I.B   Determine the value of a number written in expanded notation

1. Find the number that is the same as 9,000 + 400 + 10 + 7.

A 90,417

B 94,107

C 9,507

D 9,417

2. Find the number that is the same as 100,000 + 1,000 + 500 + 50 + 6.

A 110,556

B 115,506

C 101,556

D 11,556

3. Which one shows 21,970?

A 2,000 + 100 + 90 + 7

B 20,000 + 1,000 + 900 + 70

C 1,000 + 200 + 90 + 7

D 200,000 + 1,000 + 900 + 70

4. Which one shows 1,766?

A 1,000 + 700 + 60 + 6

B 1,000 + 700 + 6 + 6

C 10,000 + 7,000 + 600 + 60

D 170 + 60 + 6

5. Which one shows 91,505?

A 90,000 + 100 + 50 + 5

B 9,000 + 100 + 50 + 5

C 90,000 + 1,000 + 500 + 5

D 90,000 + 100 + 50 + 5

6. Find the number that is the same as 50,000 + 1,000 + 600 + 9.

A 50,169

B 51,690

C 5,169

D 51,609

7. Find the number that is the same as 10,000 + 7,000 + 600 + 50.

A 17,650

B 1,765

C 10,765

D 17,065

8. Find the number that is the same as 10,000 + 1,000 + 800 + 40 + 8.

A 10,848

B 11,848

C 18,148

D 18,848

Practice 1.C1

1. Which group has $\frac{1}{2}$ shaded?

   A

   B

   C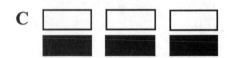

   D

2. Which group shows the fractions in order **from least to greatest**?

   A $\frac{4}{5}$ $\frac{2}{3}$ $\frac{1}{2}$ $\frac{1}{4}$

   B $\frac{1}{2}$ $\frac{1}{4}$ $\frac{2}{3}$ $\frac{4}{5}$

   C $\frac{1}{4}$ $\frac{2}{3}$ $\frac{1}{2}$ $\frac{4}{5}$

   D $\frac{1}{4}$ $\frac{1}{2}$ $\frac{2}{3}$ $\frac{4}{5}$

3. Which group shows the fractions in order **from greatest to least**?

   A $\frac{4}{7}$ $\frac{5}{6}$ $\frac{1}{5}$ $\frac{1}{3}$

   B $\frac{5}{6}$ $\frac{4}{7}$ $\frac{1}{3}$ $\frac{1}{5}$

   C $\frac{1}{5}$ $\frac{4}{7}$ $\frac{1}{3}$ $\frac{5}{6}$

   D $\frac{5}{6}$ $\frac{1}{5}$ $\frac{4}{7}$ $\frac{1}{3}$

4. Which group shows the decimals in order **from greatest to least**?

   A 0.14   0.104   0.41   0.4

   B 0.104   0.41   0.14   0.4

   C 0.41   0.4   0.14   0.104

   D 0.4   0.14   0.104   0.41

5. The figures are shaded to show equivalent fractions.

   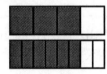

   Which is equivalent to $\frac{6}{8}$?

   A $\frac{3}{4}$

   B $\frac{1}{4}$

   C $\frac{2}{4}$

   D $\frac{5}{6}$

6. Which decimal is equal to three hundredths?

   A 0.3

   B 0.13

   C 0.03

   D 0.003

7. Which of these is true?

   A 1.04 > 1.4

   B 0.14 < 0.114

   C 0.14 < 0.041

   D 0.114 < 0.41

Practice 1.C2

I.C   Read, write, compare, and order fractions [like and unlike denominators] and decimals [to thousandths]

1. Which shows 0.12 shaded?

   A 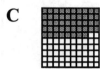   C

   B   D

2. Which group shows the fractions in order **from least to greatest**?

   A $\frac{3}{5}$ $\frac{3}{7}$ $\frac{3}{4}$ $\frac{3}{8}$

   B $\frac{3}{8}$ $\frac{3}{7}$ $\frac{3}{5}$ $\frac{3}{4}$

   C $\frac{3}{4}$ $\frac{3}{8}$ $\frac{3}{7}$ $\frac{3}{5}$

   D $\frac{3}{7}$ $\frac{3}{4}$ $\frac{3}{8}$ $\frac{3}{5}$

3. Which group shows the fractions in order **from greatest to least**?

   A $\frac{1}{3}$ $\frac{3}{5}$ $\frac{1}{6}$ $\frac{3}{4}$

   B $\frac{1}{6}$ $\frac{3}{4}$ $\frac{1}{3}$ $\frac{3}{5}$

   C $\frac{3}{5}$ $\frac{3}{4}$ $\frac{1}{6}$ $\frac{1}{3}$

   D $\frac{3}{4}$ $\frac{3}{5}$ $\frac{1}{3}$ $\frac{1}{6}$

4. The decimal 0.14 is read—

   A one and four hundredths

   B fourteen tenths

   C fourteen hundredths

   D one and four tenths

5. The figures are shaded to show two fractions.

   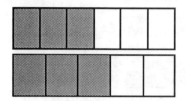

   The model shows that—

   A $\frac{3}{6} = \frac{3}{5}$

   B $\frac{3}{6} = \frac{2}{5}$

   C $\frac{3}{6} < \frac{3}{5}$

   D $\frac{3}{6} > \frac{3}{5}$

6. Which group shows the decimals in order **from greatest to least**?

   A 0.28   0.218   0.208   0.028

   B 0.208   0.218   0.028   0.28

   C 0.218   0.28   0.028   0.208

   D 0.028   0.218   0.28   0.208

7. Which of these is true?

   A $\frac{1}{4} > \frac{1}{2}$

   B $\frac{2}{3} < \frac{1}{4}$

   C $\frac{1}{3} < \frac{1}{5}$

   D $\frac{3}{4} > \frac{1}{2}$

Practice 1.C3

I.C   Read, write, compare, and order fractions [like and unlike denominators] and decimals [to thousandths]

1. Which shows $\frac{2}{3}$ shaded?

   A
   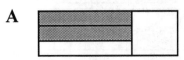

   B
   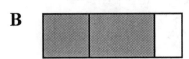

   C
   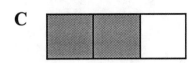

   D
   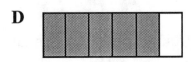

2. Which group shows the fractions in order **from least to greatest**?

   A $\frac{1}{3}$ $\frac{3}{4}$ $\frac{1}{2}$ $\frac{1}{4}$

   B $\frac{1}{2}$ $\frac{1}{3}$ $\frac{1}{4}$ $\frac{3}{4}$

   C $\frac{1}{4}$ $\frac{1}{2}$ $\frac{3}{4}$ $\frac{1}{3}$

   D $\frac{1}{4}$ $\frac{1}{3}$ $\frac{1}{2}$ $\frac{3}{4}$

3. Which group shows the fractions in order **from greatest to least**?

   A $\frac{2}{3}$ $\frac{4}{5}$ $\frac{3}{4}$ $\frac{5}{6}$

   B $\frac{5}{6}$ $\frac{4}{5}$ $\frac{3}{4}$ $\frac{2}{3}$

   C $\frac{4}{5}$ $\frac{5}{6}$ $\frac{2}{3}$ $\frac{3}{4}$

   D $\frac{3}{4}$ $\frac{2}{3}$ $\frac{5}{6}$ $\frac{4}{5}$

4. Which group shows the decimals in order **from greatest to least**?

   A 1.1   1.01   1.15   1.2

   B 1.01   1.15   1.1   1.2

   C 1.15   1.01   1.1   1.2

   D 1.2   1.15   1.1   1.01

5. The figures are shaded to show two fractions.

   The model shows that—

   A $\frac{3}{7} = \frac{1}{2}$

   B $\frac{4}{7} > \frac{1}{2}$

   C $\frac{4}{7} < \frac{1}{3}$

   D $\frac{1}{2} = \frac{4}{7}$

6. The decimal 0.07 is read—

   A seven

   B seven tenths

   C seven hundredths

   D seventh thousandths

7. Which of these is true?

   A 0.8 < 0.15

   B 0.18 < 0.09

   C 0.1 < 0.02

   D 0.5 > 0.16

Practice 1.D1

I.D Name and write mixed numbers as improper fractions and improper fractions as mixed numbers

1. Which improper fraction equals $2\frac{3}{4}$?

A $\frac{9}{4}$

B $\frac{5}{4}$

C $\frac{11}{4}$

D $\frac{8}{4}$

2. Which improper fraction equals $1\frac{3}{5}$?

A $\frac{8}{5}$

B $\frac{9}{5}$

C $\frac{4}{5}$

D $\frac{5}{5}$

3. Which mixed number represents the shaded parts of the boxes?

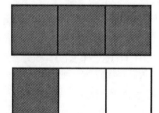

A $3\frac{1}{3}$

B $2\frac{1}{3}$

C $1\frac{1}{2}$

D $1\frac{1}{3}$

4. Which improper fraction represents the shaded parts of the boxes?

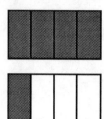

A $\frac{5}{4}$

B $\frac{3}{4}$

C $\frac{5}{2}$

D $\frac{8}{4}$

5. Which mixed number equals $\frac{7}{2}$?

A $4\frac{1}{2}$

B $3\frac{1}{2}$

C $7\frac{1}{2}$

D $4\frac{1}{3}$

6. Which mixed number equals $\frac{4}{3}$?

A $1\frac{1}{2}$

B $2\frac{2}{3}$

C $4\frac{1}{3}$

D $1\frac{1}{3}$

Practice 1.D2

I.D *Name and write mixed numbers as improper fractions and improper fractions as mixed numbers*

1. Which improper fraction equals $4\frac{1}{4}$?

   A $\frac{4}{4}$

   B $\frac{17}{4}$

   C $\frac{8}{4}$

   D $\frac{5}{4}$

2. Which improper fraction equals $3\frac{1}{5}$?

   A $\frac{15}{5}$

   B $\frac{4}{5}$

   C $\frac{8}{5}$

   D $\frac{16}{5}$

3. Which mixed number represents the shaded parts of the boxes?

   A $1\frac{1}{2}$

   B $1\frac{3}{4}$

   C $1\frac{2}{3}$

   D $1\frac{1}{4}$

4. Which improper fraction represents the shaded parts of the boxes?

   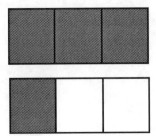

   A $\frac{6}{3}$

   B $\frac{4}{2}$

   C $\frac{2}{3}$

   D $\frac{4}{3}$

5. Which mixed number equals $\frac{6}{5}$?

   A $2\frac{3}{5}$

   B $1\frac{2}{5}$

   C $1\frac{1}{5}$

   D $6\frac{1}{5}$

6. Which mixed number equals $\frac{16}{3}$?

   A $2\frac{1}{3}$

   B $6\frac{1}{3}$

   C $5\frac{1}{3}$

   D $4\frac{2}{3}$

Practice 1.D3

I.D   Name and write mixed numbers as improper fractions and improper fractions as mixed numbers

1. Which improper fraction equal $1\frac{2}{3}$?

   A $\frac{5}{3}$

   B $\frac{3}{3}$

   C $\frac{6}{3}$

   D $\frac{10}{3}$

2. Which improper fraction equals $2\frac{1}{6}$?

   A $\frac{8}{6}$

   B $\frac{9}{6}$

   C $\frac{13}{6}$

   D $\frac{12}{6}$

3. Which mixed number represents the shaded parts of the boxes?

   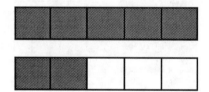

   A $5\frac{2}{5}$

   B $1\frac{3}{5}$

   C $1\frac{2}{5}$

   D $1\frac{1}{3}$

4. Which improper fraction represents the shaded parts of the boxes?

   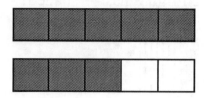

   A $\frac{2}{5}$

   B $\frac{8}{5}$

   C $\frac{7}{5}$

   D $\frac{8}{2}$

5. Which mixed number equals $\frac{11}{5}$?

   A $3\frac{1}{5}$

   B $2\frac{1}{5}$

   C $4\frac{1}{5}$

   D $1\frac{4}{5}$

6. Which improper fraction equals $1\frac{2}{3}$?

   A $\frac{5}{2}$

   B $\frac{6}{3}$

   C $\frac{5}{3}$

   D $\frac{3}{3}$

Practice 1.E1

1. Which figure has $\frac{3}{10}$ shaded?

A

B

C

D

2. Which decimal equals $\frac{16}{100}$?

   A 1.006

   B 0.016

   C 1.6

   D 0.16

3. Which decimal equals $\frac{45}{100}$?

   A 0.0045

   B 0.045

   C 0.45

   D 4.005

4. Which decimal tells how much of the square is shaded?

   A 0.16

   B 0.06

   C 0.6

   D 1.6

5. Which fraction tells how much of the figure is shaded?

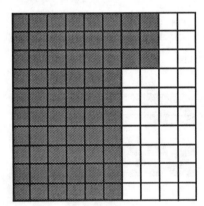

   A $\frac{6}{100}$

   B $\frac{16}{100}$

   C $\frac{60}{100}$

   D $\frac{66}{100}$

Practice 1.E2

I.E Relate decimals to fractions that name tenths and hundredths [with and without models]

1. Which decimal tells how much of the squares are shaded?

  A 0.108

  B 0.18

  C 1.08

  D 1.8

2. Which fraction is equal to 0.27?

  A $\dfrac{2}{7}$

  B $\dfrac{27}{100}$

  C $\dfrac{27}{1000}$

  D $\dfrac{27}{10}$

3. Which fraction is equal to 0.55?

  A $\dfrac{5}{10}$

  B $\dfrac{1}{2}$

  C $\dfrac{55}{100}$

  D $\dfrac{55}{10}$

4. Which fraction tells how much of the square is shaded?

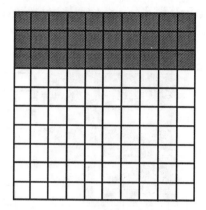

  A $\dfrac{3}{100}$

  B $\dfrac{13}{100}$

  C $\dfrac{30}{100}$

  D $\dfrac{40}{100}$

5. Which decimal is equal to $\dfrac{81}{100}$?

  A 8.1

  B 8.01

  C 0.081

  D 0.81

6. Which decimal is equal to $\dfrac{9}{10}$?

  A 0.09

  B 0.9

  C 0.19

  D 1.9

Practice 1.E3

I.E   Relate decimals to fractions that name tenths and hundredths [with and without models]

1. Which decimal tells how much of the square is shaded?

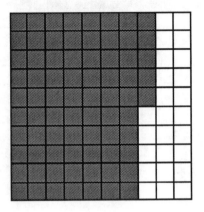

    A 0.57

    B 0.075

    C 7.5

    D 0.75

2. Which fraction is equal to 0.4?

    A $\dfrac{1}{5}$

    B $\dfrac{4}{10}$

    C $\dfrac{4}{100}$

    D $\dfrac{1}{4}$

3. Which decimal is equal to $\dfrac{5}{10}$?

    A 0.05

    B 0.15

    C 0.5

    D 0.105

4. Which fraction tells how much of the square is shaded?

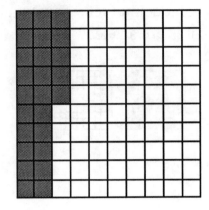

    A $\dfrac{25}{100}$

    B $\dfrac{20}{100}$

    C $\dfrac{1}{25}$

    D $\dfrac{4}{100}$

5. Which decimal equals $\dfrac{12}{100}$?

    A 1.2

    B 0.12

    C 0.012

    D 1.02

6. Which fraction is equal to 0.05?

    A $\dfrac{5}{10}$

    B $\dfrac{15}{100}$

    C $\dfrac{50}{100}$

    D $\dfrac{5}{100}$

Mathematical Relations, Functions, & Algebraic Concepts

II. Demonstrate an understanding of mathematical relations, functions, and other algebraic concepts

   A. Identify and extend whole-number and geometric patterns
   B. Identify number sentences that show the inverse relationships between addition/subtraction and multiplication/division (fact families)
   C. Identify the relationship between two sets of related data, such as ordered pairs in a table

Notes

Objective 2: Pretest

II.A Identify and extend whole-number and geometric patterns (1-5)

1. Which number goes in the empty space in this number pattern?

15, 21, 27, _____, 39, 45

A 29

B 30

C 32

D 33

2. Which number goes in the empty space of this number pattern?

42, 39, 36, _____, 30, 27

A 32

B 33

C 34

D 38

3. Dora put books on her bookshelves. She put 6 books on the first shelf, 10 books on the second shelf, and 14 books on the third shelf. If she continues the pattern, how many books would she put on the **fifth** shelf?

A 24

B 22

C 20

D 18

4. Which figure goes in the empty space?

A ○

B ◁

C ▭

D ☐

5. Which figure goes in the empty space?

● ○ ◁ ● ○ ▭ ● __ ◁

A ○

B ▭

C ◁

D ●

6. Which number sentence belongs to the same family of facts as $8 + 5 = 13$?

   A $10 + 3 = 13$

   B $13 - 8 = 5$

   C $8 \times 5 = 40$

   D $13 + 8 = 21$

7. Which number sentence belongs to the same family of facts as $7 \times 4 = 28$?

   A $7 + 4 = 11$

   B $7 - 4 = 3$

   C $28 \div 7 = 4$

   D $14 + 14 = 28$

8. Which number sentence belongs to the same family of facts as $18 - 9 = 9$?

   A $9 - 9 = 0$

   B $18 + 9 = 27$

   C $9 \times 9 = 81$

   D $9 + 9 = 18$

9. Which number sentence belongs to the same family of facts as $42 \div 7 = 6$?

   A $7 - 6 = 1$

   B $7 \times 6 = 42$

   C $42 + 7 = 49$

   D $7 + 6 = 13$

10. Danni had 45 pieces of candy. She put 3 pieces of candy in each party bag she made. Which expression shows how to find the number of party bags Danni made?

    A $3 \div 45$

    B $3 + 45$

    C 3×45

    D $45 \div 3$

11. There are 35 chairs in Mr. Ward's classroom. He wants to put the chairs in 7 rows, and he wants the same number of chairs in each row. Which expression shows how to find the number of chairs he should put in each row?

    A $35 + 7$

    B $35 - 7$

    C $35 \div 7$

    D 35×7

12. There are 24 students in Mr. Reyna's class. On a field trip, 12 of the students rode in the school van. Which expression shows how to find the number of students that did not ride in the school van?

A 12 + 24

B 24 − 12

C 24 ÷ 12

D 24 x 12

13. Ricky began with 20 marbles, but lost 4 of them in a game. Which expression shows how to find the number of marbles he had left?

A 20 − 4

B 20 ÷ 4

C 20 x 4

D 20 + 4

II.C *Identify the relationship between two sets of related data, such as ordered pairs in a table (14–18)*

14. For a card game, Sean dealt cards in a pattern. The chart shows how many cards he gave to the players.

Player	Cards
1	2
2	4
3	
4	8
5	10

How many cards did player 3 get?

A 4

B 5

C 6

D 10

15. Terry likes to make bracelets for her friends. For 1 bracelet, she uses 3 large beads. For 2 bracelets, she uses 6 large beads. For 3 bracelets, she uses 9 large beads. How many large beads does she use for 4 bracelets?

A 9

B 10

C 12

D 14

16. At a pet store, the clerk uses a pattern to place fish in different tanks. The chart shows the number of fish in each tank.

Tank	Fish
1	5
2	10
3	15
4	
5	25

How many fish will the clerk put in tank 4?

A 18

B 20

C 23

D 30

17. In a race, the first-place winner got 20 tokens. The second-place winner got 14 tokens. The fourth-place winner got 2 tokens. How many tokens did the **third-place** winner get?

A 7

B 8

C 9

D 12

18. At a repair shop, 3 workers fix bicycles. The chart shows how many minutes it takes them to fix different numbers of bicycles.

Bicycles	Minutes
3	30
	60
9	90
12	120
15	150

How many bicycles can the workers fix in 60 minutes?

A 2

B 4

C 6

D 10

Practice 2.A1

II.A Identify and extend whole-number and geometric patterns

1. Which number goes in the empty space of this number pattern?

   8, 12, 16, ____, 24, 28

   A 18
   B 20
   C 22
   D 25

2. Which number goes in the empty space of this number pattern?

   90, 85, 80, 75, ____, 65

   A 60
   B 68
   C 70
   D 75

3. Mike arranged his soccer trophies in a pattern. He placed 2 trophies in the first row, 4 trophies in the second row, and 6 trophies in the third row. If he continues the pattern, how many trophies will he put in the **fifth** row?

   A 6
   B 8
   C 9
   D 10

4. Which figure goes in the empty space?

   A ☐

   B ☐

   C ☐

   D ○

5. Which figure goes in the empty space?

   A ◎

   B ☐

   C ◁

   D △

Practice 2.A2

II.A *Identify and extend whole-number and geometric patterns*

1. Which number is missing in the number pattern?

   13, 17, 21, 25, ____, 33, 37

   A 32
   B 30
   C 29
   D 27

2. What is the missing number in the number pattern?

   3, 6, 12, ____, 48, 96

   A 15
   B 18
   C 20
   D 24

3. At a football game, there are 3 cheerleaders in the first row, 6 cheerleaders in the second row, and 9 cheerleaders in the third row. If the pattern continues, how many cheerleaders will be in the **fifth** row?

   A 12
   B 15
   C 17
   D 20

4. Which figure goes in the empty space?

   A
   B
   C
   D

5. Which figure goes in the empty space?

   A
   B
   C
   D

Practice 2.A3

II.A   Identify and extend whole-number and geometric patterns

1. Which number is missing in the number pattern?

23, 28, 33, 38, _____, 48, 53 …

A 46

B 45

C 44

D 43

2. What is the missing number in the number pattern?

2, 4, 8, 16, 32, _____, 128 …

A 48

B 56

C 64

D 112

3. Joseph made a design with stars. He put 4 stars in the first row, 7 stars in the second row, and 10 stars in the third row. If he continued the pattern, how many stars did Joseph use in the **sixth** row?

A 13

B 15

C 16

D 19

4. Which figure goes in the empty space?

A ▭

B ▢

C ○

D ○

5. Which figure goes in the empty space?

A

B

C

D

Practice 2.B1

II.B Identify number sentences that show the inverse relationships between addition/subtraction and multiplication/division (fact families)

1. After Grace won 8 pieces of candy in a math game, she had 42 pieces of candy. Which expression shows how to find the number of candies she had **before** winning the math game?

   A $42 - 8$

   B $42 + 8$

   C 42×8

   D $42 \div 8$

2. If 125 minus a number equals 76, which expression shows how to find the number?

   A $125 + 76$

   B 125×76

   C $125 - 76$

   D $125 \div 76$

3. Which number sentence belongs to the same family of facts as $56 \div 8 = 7$?

   A $56 - 8 = 48$

   B $8 + 7 = 15$

   C $7 \times 8 = 56$

   D $56 + 7 = 63$

4. Which number sentence belongs to the same family of facts as $9 \times 4 = 36$?

   A $36 \div 4 = 9$

   B $36 \times 4 = 144$

   C $9 + 4 = 13$

   D $36 + 4 = 40$

5. In a band, 65 members march in rows. In each row, there are 5 band members. Which expression shows how to find how many rows of band members there are?

   A 65×5

   B $65 - 5$

   C $65 + 5$

   D $65 \div 5$

6. Which number sentence belongs to the same family of facts as $7 + 6 = 13$?

   A $13 + 7 = 20$

   B $13 - 7 = 6$

   C $7 - 6 = 1$

   D $10 + 3 = 13$

Practice 2.B2

II.B *Identify number sentences that show the inverse relationships between addition/subtraction and multiplication/division (fact families)*

1. Which number sentence belongs to the same family of facts as $16 - 8 = 8$?

   A $8 - 8 = 0$

   B $9 + 7 = 16$

   C $16 + 8 = 24$

   D $8 + 8 = 16$

2. Mrs. Franz made 7 small pies. She cut a total of 28 equal pieces from all 7 pies. Which expression shows how to find how many pieces she cut from each pie?

   A $28 \div 7$

   B 28×7

   C $28 - 7$

   D $28 + 7$

3. Which number sentence belongs to the same family of facts as $21 \div 7 = 3$?

   A $7 - 3 = 4$

   B $21 + 7 = 28$

   C $7 \times 3 = 21$

   D $21 + 3 = 24$

4. Clark used 36 flowers to make 9 party decorations. He used the same number of flowers in each decoration. Which expression shows how to find the number of flowers he used in each decoration?

   A $36 \div 9$

   B 36×9

   C $36 + 9$

   D $36 - 9$

5. Which number sentence belongs to the same family of facts as $8 \times 6 = 48$?

   A $48 - 8 = 40$

   B $48 \div 8 = 6$

   C $8 + 6 = 14$

   D $48 + 6 = 54$

6. Which number sentence belongs to the same family of facts as $7 + 6 = 13$?

   A $13 + 7 = 20$

   B $13 - 7 = 6$

   C $7 - 6 = 1$

   D $10 + 3 = 13$

Practice 2.B3

II.B *Identify number sentences that show the inverse relationships between addition/subtraction and multiplication/division (fact families)*

1. By the end of the school day, 27 students had signed up for a field trip. Only 9 students signed up before lunch. Which expression shows how to find the number of students who signed up after lunch?

   A 27 + 9

   B 27 x 9

   C 27 ÷ 9

   D 27 − 9

2. Eliza used 24 buttons when she made 6 doll dresses. She used the same number of buttons on each dress. Which expression shows how to find the number of buttons she used on each dress?

   A 24 − 6

   B 24 x 6

   C 24 ÷ 6

   D 24 + 6

3. Which number sentence belongs to the same family of facts as 6 x 3 = 18?

   A 18 ÷ 3 = 6

   B 9 + 9 = 18

   C 3 x 18 = 54

   D 18 ÷ 9 = 2

4. After loaning her sister $5, Missy had $15 left. Which expression shows how to find the amount of money Missy had before loaning any to her sister?

   A 15 ÷ 5

   B 15 x 5

   C 15 + 5

   D 15 − 5

5. Which number sentence belongs to the same family of facts as 42 ÷ 6 = 7?

   A 42 − 7 = 35

   B 7 x 6 = 42

   C 7 + 6 = 13

   D 42 + 6 = 48

6. A math test had 45 problems. Rashid skipped 3 problems because he ran out of time. Which expression shows how to find the number of problems Rashid completed?

   A 45 + 3

   B 45 ÷ 3

   C 45 − 3

   D 45 x 3

Practice 2.C1

II.C Identify the relationship between two sets of related data, such as ordered pairs in a table

1. Mrs. Peterson followed a pattern to put students in groups. The chart shows how many students she put in each group.

Group	Students
1	5
2	
3	9
4	11

How many students did Group 2 have?

A 6

B 7

C 8

D 9

2. In a race, the first-place winner won $250. The second-place winner won $200. The fourth-place winner won $100. How much money did the **third**-place winner get?

A $50

B $100

C $125

D $150

3. Jesse wanted to give his old toys away to different charities. He followed a pattern to put them in piles. The chart shows how many toys he put in each pile.

Charity	Toys
1	5
2	10
3	
4	20

How many toys did Charity 3 get?

A 10

B 12

C 13

D 15

4. Allen mows lawns to earn money in the summer. For 1 lawn, he earns $12. For 2 lawns, he earns $24. For 3 lawns he earns $36. How much will he earn for mowing **four** lawns?

A $36

B $40

C $48

D $58

Practice 2.C2

II.C Identify the relationship between two sets of related data, such as ordered pairs in a table

1. At a bakery, a baker makes muffins in the morning. The chart shows how many minutes it takes the baker to make different numbers of muffins.

Muffins	Time
12	10 min
24	25 min
48	40 min
	55 min
192	70 min

How many muffins can the baker make in 55 minutes?

A 88

B 90

C 96

D 192

2. Jeannie may watch 1 hour of TV if she spends 2 hours on homework. She may watch 2 hours if she spends 4 hours on homework. She may watch 4 hours if she spends 8 hours on homework. How many hours of TV may she watch if she spends 6 hours on her homework?

A 2

B 4

C 5

D 3

3. The first 5 people who entered the carnival gates won tickets for a raffle. The chart shows how many tickets each person won.

Person	Tickets
1st	10
2nd	8
3rd	6
4th	
5th	2

How many tickets did the 4th person win?

A 6

B 4

C 3

D 2

4. At the flower shop, Mrs. Edwards follows a pattern when she makes gift vases. She uses 2 roses in the small vase. She uses 6 roses in the medium vase and 10 roses in the large vase. How many roses does she probably use in the extra large vase?

A 14

B 12

C 10

D 8

Practice 2.C3

II.C Identify the relationship between two sets of related data, such as ordered pairs in a table

1. A store owner gives bonus points to customers when they spend a certain amount of money. The chart shows how many bonus points a customer can earn for spending from $10 to $50.

$ Spent	Points
$10	1
$20	2
$30	4
$40	8
$50	

How many bonus points would a customer earn for spending $50?

A 12

B 14

C 16

D 20

2. Zoe drinks 16 ounces of water after running 1 mile. She drinks 32 ounces after running 2 miles and 128 ounces after running 4 miles. How much water does she probably drink after running 3 miles?

A 40 oz

B 48 oz

C 56 oz

D 64 oz

3. The chart shows how much vacation time Chuck earns for each month he works.

Months Worked	Vacation Time Earned
1	0.5 day
2	1 day
3	1.5 days
4	2 days
5	

How much vacation time will Chuck earn after working for 5 months?

A 2.5 days

B 3.5 days

C 3 days

D 4 days

4. Xavier works at a candy factory. He can make 1 batch of candy in 15 minutes, 2 batches of candy in 30 minutes, and 3 batches of candy in 45 minutes. How many minutes will it take him to make 4 batches of candy?

A 65 minutes

B 60 minutes

C 50 minutes

D 40 minutes

Notes

Geometric Properties/ Relationships

III. Demonstrate an understanding of geometric properties and relationships

A. Identify right, acute, obtuse, and straight angles
B. Identify models of intersecting, parallel, and perpendicular lines
C. Identify parts of a circle (center, radius, chord, diameter)
D. Identify and describe shapes and solids in terms of their properties (sides, vertices, edges, faces)
E. Identify congruent shapes
F. Identify lines of symmetry in shapes
G. Locate and name points on a number line using whole numbers, fractions (halves and fourths), and decimals (tenths)

Notes

Objective 3: Pretest

III.A Identify right, acute, obtuse, and straight angles (1-6)

1. Which one shows a picture of an acute angle?

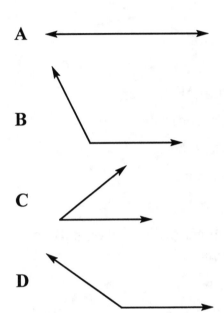

A

B

C

D

2. Look at the angle below. What kind of angle is it?

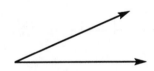

   A straight

   B obtuse

   C right

   D acute

3. Which figure appears to have at least one right angle?

A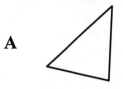

B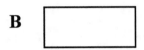

C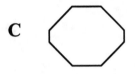

D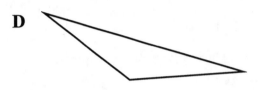

4. Look at the angle below. What kind of angle is it?

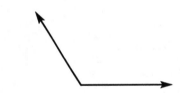

   A obtuse

   B acute

   C right

   D straight

5. Which figure has 3 acute angles?

A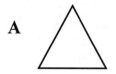

B

C

D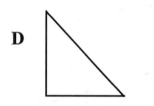

6. Look at the angle below. What kind of angle is it?

A obtuse
B right
C acute
D straight

III.B *Identify models of intersecting, parallel, and perpendicular lines (7-12)*

7. What are these lines called?

A parallel
B perpendicular
C intersecting
D horizontal

8. Which set of lines seems to be parallel?

A

B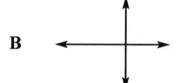

C

D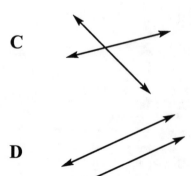

9. Which figure has at least one set of parallel lines?

A

B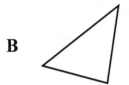

C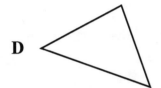

D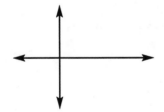

11. Which set of line segments is intersecting?

A

B

C

D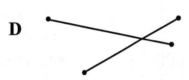

10. What are these lines called?

A parallel

B perpendicular

C horizontal

D vertical

12. What are these lines called?

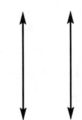

A intersecting

B perpendicular

C parallel

D horizontal

III.C *Identify the parts of a circle [center, radius, chord, diameter] (13-18)*

Use the circle to answer questions 13-15.

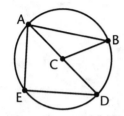

13. Which of the following is a chord?

    A $\overline{BC}$

    B $\overline{AD}$

    C $\overline{CD}$

    D $\overline{DE}$

14. Which of these passes through the center of the circle?

    A $\overline{DE}$

    B $\overline{AB}$

    C $\overline{AD}$

    D $\overline{AE}$

15. Which of these is a radius?

    A $\overline{BC}$

    B $\overline{AD}$

    C $\overline{DE}$

    D $\overline{AB}$

Use the circle to answer questions 16-18.

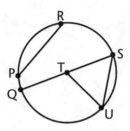

16. $\overline{TU}$ is a—

    A diameter

    B radius

    C chord

    D parallel

17. The diameter of the circle is—

    A $\overline{PR}$

    B $\overline{ST}$

    C $\overline{QT}$

    D $\overline{QS}$

18. Which of the following is a chord?

    A $\overline{QT}$

    B $\overline{ST}$

    C $\overline{SU}$

    D $\overline{QS}$

III.D Identify and describe shapes and solids in terms of their properties [sides, vertices, edges, faces] (19-24)

19. Which figure represents a hexagon?

A

B

C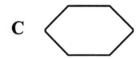

D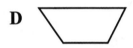

20. The box has the shape of a—

  A pyramid

  B sphere

  C cone

  D cube

21. How many sides does a pentagon have?

  A 3

  B 4

  C 5

  D 6

22. Which figure represents a cone?

A

B

C

D

23. How many edges does this figure have?

  A 12

  B 10

  C 8

  D 6

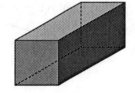

24. A basketball has the shape of a—

  A quadrilateral

  B cube

  C cylinder

  D sphere

III.E *Identify congruent shapes (25-28)*

25. Which pair of triangles appear to be congruent figures?

A

B

C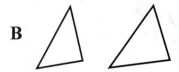

D

26. Which figure appears to be congruent with this figure?

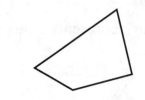

A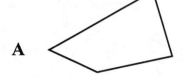

B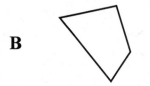

C

D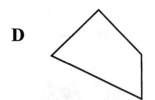

27. Which one shows a pair of congruent figures?

A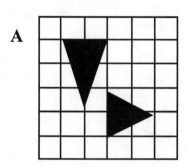

B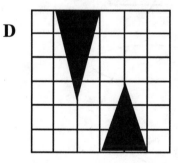

C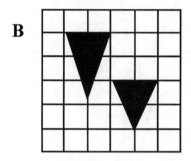

D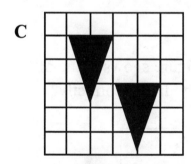

28. Which two parts of the square appear to be congruent?

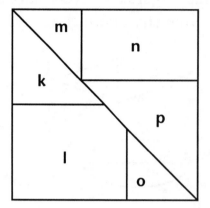

A l and n

B m and p

C k and p

D m and o

III.F *Identify lines of symmetry in shapes (29-32)*

29. Which figure does **NOT** have a line of symmetry drawn correctly?

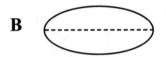

A

B

C

D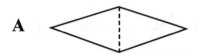

30. Which shape does not have a line of symmetry?

A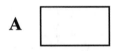

B

C

D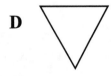

31. Which figure has more than one line of symmetry?

A

B

C

D

32. Which figure shows a line of symmetry drawn correctly?

A

B

C

D

III.G Locate and name points on a number line using whole numbers, fractions [halves and fourths], and decimals [tenths] (33-38)

33. Which point best represents 6.2 on the number line?

A L

B M

C N

D O

34. Which number line shows all the whole numbers that are greater than 2 and less than 6?

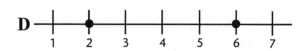

35. Which point best represents $10\frac{1}{2}$ on the number line?

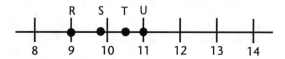

A U

B T

C S

D R

36. On which number line does the arrow point most closely to 7.8?

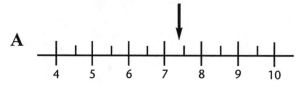

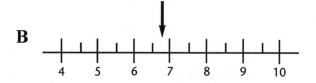

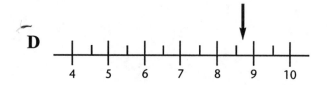

37. What number does point R best represent?

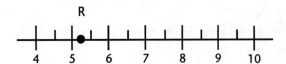

A $5\frac{3}{4}$

B $5\frac{1}{2}$

C $5\frac{1}{4}$

D 5

38. Which number line shows whole numbers greater than 8?

A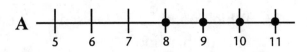

B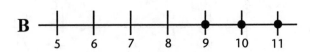

C

D

Practice 3.A1

III.A Identify right, acute, obtuse, and straight angles

1. Which one shows a picture of an obtuse angle?

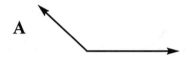

A

B

C

D

2. Look at the angle below. What kind of angle is it?

A straight

B right

C obtuse

D acute

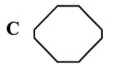

3. Look at the angle below. What kind of angle is it?

A obtuse

B acute

C right

D straight

4. Which figure appears to have at least one right angle?

A

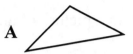

B

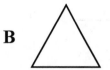

C

D

5. Which figure appears to have at least two obtuse angles?

A

B

C

D

Practice 3.A2

1. Which one shows a picture of an acute angle?

   A

   B

   C

   D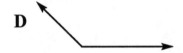

2. Look at the angle below. What kind of angle is it?

   A straight

   B acute

   C obtuse

   D right

3. Which figure appears to have three acute angles?

   A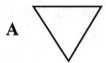

   B

   C

   D

4. Look at the angle below. What kind of angle is it?

   A straight

   B acute

   C right

   D obtuse

5. Which figure appears to have at least one right angle?

   A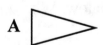

   B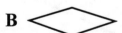

   C

   D

Practice 3.A3

III.A Identify right, acute, obtuse, and straight angles

1. Which one shows a picture of a right angle?

   A

   B

   C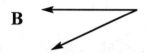

   D

2. Look at the angle below. What kind of angle is it?

   A right
   B straight
   C acute
   D obtuse

3. Which figure appears to have one obtuse angle?

   A

   B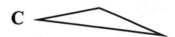

   C

   D

4. Look at the angle below. What kind of angle is it?

   A straight
   B acute
   C right
   D obtuse

5. Look at the angle below. What kind of angle is it?

   A right
   B obtuse
   C acute
   D straight

Practice 3.B1

1. What are these lines called?

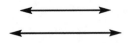

   A parallel

   B perpendicular

   C intersecting

   D vertical

2. Which set of lines seems to be perpendicular?

   A

   B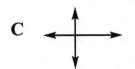

   C

   D

3. What are these lines called?

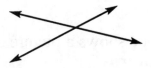

   A perpendicular

   B parallel

   C intersecting

   D horizontal

4. Which figure has at least one set of parallel lines?

   A

   B

   C

   D

5. Look at the lines below. They are called—

   A vertical

   B parallel

   C simple

   D perpendicular

Practice 3.B2

III.B Identify models of intersecting, parallel, and perpendicular lines

1. What are these lines called?

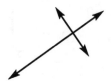

    A parallel

    B simple

    C perpendicular

    D horizontal

2. Which set of lines seems to be parallel?

    A

    B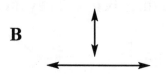

    C

    D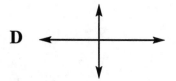

3. Look at the lines below. They are called—

    A parallel

    B vertical

    C perpendicular

    D intersecting

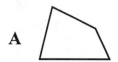

4. Which figure has at least one set of perpendicular lines?

    A

    B

    C

    D

5. Look at the lines below. They are called—

    A horizontal

    B parallel

    C intersecting

    D perpendicular

Practice 3.B3

III.B Identify models of intersecting, parallel, and perpendicular lines

1. What are these lines called?

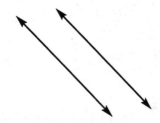

   A vertical

   B perpendicular

   C intersecting

   D parallel

2. Which set of lines seems to be perpendicular?

   A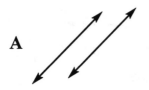

   B

   C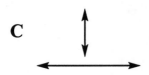

   D

3. Look at the figure below. $\overline{AB}$ and $\overline{BC}$ appear to be—

   A parallel

   B crossing

   C perpendicular

   D horizontal

4. Which figure has at least one set of parallel line segments?

   A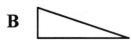

   B

   C

   D

5. Look at the lines below. They are called—

   A intersecting

   B parallel

   C vertical

   D perpendicular

   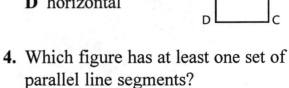

Practice 3.C1

III.C Identify the parts of a circle (center, radius, chord, diameter)

Use the circle to answer questions 1-3.

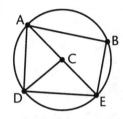

1. Which of the following is a chord?

   A $\overline{AC}$

   B $\overline{AE}$

   C $\overline{CD}$

   D $\overline{DE}$

2. Which of the following is a radius?

   A $\overline{AE}$

   B $\overline{AB}$

   C $\overline{CE}$

   D $\overline{DE}$

3. Which of these passes through the center of the circle?

   A $\overline{CD}$

   B $\overline{AE}$

   C $\overline{AB}$

   D $\overline{DE}$

Use the circle to answer questions 4-6.

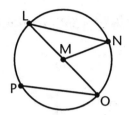

4. The diameter of the circle is—

   A $\overline{OP}$

   B $\overline{LM}$

   C $\overline{LO}$

   D $\overline{MN}$

5. $\overline{MN}$ is a—

   A radius

   B diameter

   C chord

   D perpendicular

6. Which of the following is a chord?

   A $\overline{MO}$

   B $\overline{LN}$

   C $\overline{LO}$

   D $\overline{MN}$

Practice 3.C2

III.C Identify the parts of a circle (center, radius, chord, diameter)

Use the circle to answer questions 1-3.

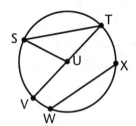

1. Which of the following is a diameter?

   A $\overline{WX}$

   B $\overline{ST}$

   C $\overline{TV}$

   D $\overline{SU}$

2. Which of the following is a chord?

   A $\overline{VU}$

   B $\overline{VT}$

   C $\overline{US}$

   D $\overline{ST}$

3. Which of these is a radius?

   A $\overline{UT}$

   B $\overline{VT}$

   C $\overline{WX}$

   D $\overline{ST}$

Use the circle to answer questions 4-6.

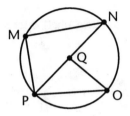

4. Which of these is a radius?

   A $\overline{MN}$

   B $\overline{NP}$

   C $\overline{OQ}$

   D $\overline{OP}$

5. Which of these passes through the center of the circle?

   A $\overline{OQ}$

   B $\overline{NP}$

   C $\overline{OP}$

   D $\overline{NQ}$

6. Which of the following is a diameter?

   A $\overline{MN}$

   B $\overline{MP}$

   C $\overline{OQ}$

   D $\overline{NP}$

Practice 3.C3

Use the circle to answer questions 1-3.

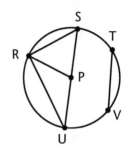

1. $\overline{RS}$ is a—

   A diameter

   B radius

   C parallel

   D chord

2. The diameter of the circle is—

   A $\overline{PR}$

   B $\overline{TV}$

   C $\overline{SU}$

   D $\overline{PU}$

3. Which of these is a radius?

   A $\overline{TV}$

   B $\overline{PS}$

   C $\overline{SU}$

   D $\overline{RU}$

Use the circle to answer questions 4-6.

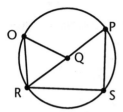

4. Which of these is **NOT** a chord?

   A $\overline{PQ}$

   B $\overline{OR}$

   C $\overline{PS}$

   D $\overline{RS}$

5. Which of the following is the diameter of the circle?

   A $\overline{OQ}$

   B $\overline{OR}$

   C $\overline{PR}$

   D $\overline{RS}$

6. Which of the following is a radius?

   A $\overline{OQ}$

   B $\overline{OR}$

   C $\overline{PR}$

   D $\overline{PS}$

Practice 3.D1

1. Look at this group of figures.

Which figure could be included in this group?

A

B

C

D

2. How many faces does this pyramid have?

- **A** 4
- **B** 5
- **C** 6
- **D** 8

3. How many sides does this figure have?

- **A** 3
- **B** 4
- **C** 5
- **D** 6

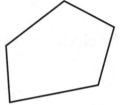

4. Which figure is **NOT** a quadrilateral?

A

B

C

D

5. How many edges does this figure have?

- **A** 3
- **B** 6
- **C** 8
- **D** 12

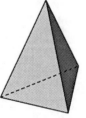

Practice 3.E1

III.E Identify congruent shapes

1. Which pair of triangles appears to be congruent figures?

A

B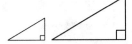

C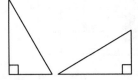

D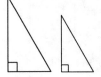

2. Which figure appears to be congruent with the figure to the right?

A

B

C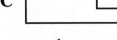

D

3. Which one shows a pair of congruent figures?

A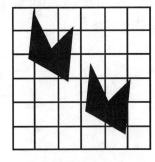

B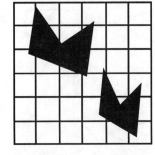

C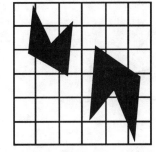

D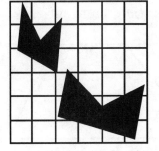

Practice 3.E2

III.E *Identify congruent shapes*

1. Which pair of shapes appear to be congruent figures?

A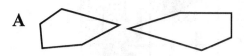

B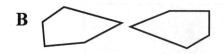

C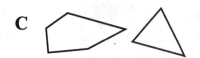

D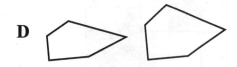

2. Which two parts of the rectangle appear to be congruent?

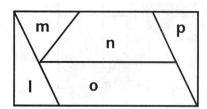

A l and m

B n and o

C p and m

D l and p

3. Which one shows a pair of congruent figures?

A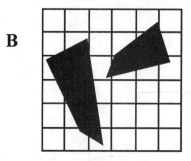

B

C

D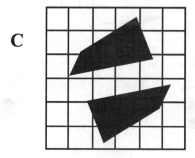

Practice 3.E3

1. Which pair of shapes appear to be congruent figures?

A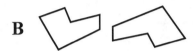

B

C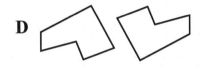

D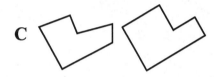

2. Which two parts of the circle appear to be congruent?

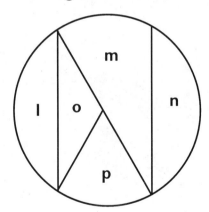

A m and o

B l and n

C p and o

D n and p

3. Which one shows a pair of congruent figures?

A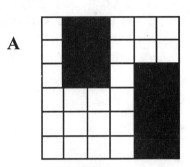

B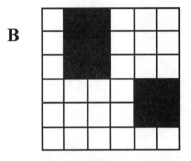

C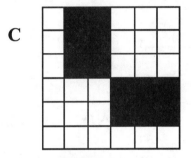

D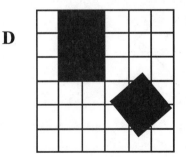

Practice 3.F1

1. Which figure has a line of symmetry drawn correctly?

A

B

C

D

2. Which shape does **NOT** have a line of symmetry?

A

B

C

D

3. Which figure has more than one line of symmetry?

A

B

C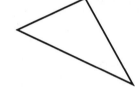

D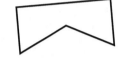

4. Which figure shows a line of symmetry drawn correctly?

A

B

C

D

Practice 3.F2

III.F Identify lines of symmetry in shapes

1. The picture shows a diagram of Miller Park and paths that cut through it. Which path is a line of symmetry through the park?

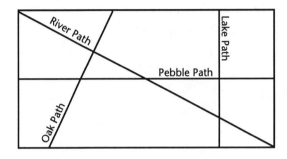

  A Oak Path
  B Lake Path
  C River Path
  D Pebble Path

2. Which shape does **NOT** have a line of symmetry?

  A

  B

  C

  D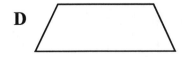

3. Which figure has only one line of symmetry?

  A

  B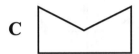

  C

  D

4. The picture shows an oval place mat. Judith wants to fold the mat along a line of symmetry. Which line shows where Judith should fold the mat?

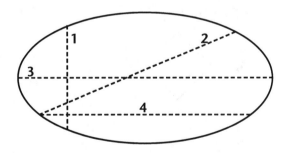

  A line 1
  B line 2
  C line 3
  D line 4

Practice 3.F3

1. The picture shows a diagram of a candy bar. Eddie wants to cut the candy bar along a line of symmetry and share it with his friend. Which line shows where Eddie should cut the candy bar?

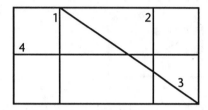

   A line 1
   B line 2
   C line 3
   D line 4

2. Which shape has more than one line of symmetry?

   A

   B

   C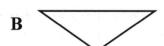

   D

3. Which figure has a line of symmetry drawn correctly?

   A

   B

   C

   D

4. Dana wants to divide her garden along a line of symmetry, growing tulips on one side of the line and roses on the other side. Which one could be a diagram of Dana's garden?

   A Tulips Roses

   B Tulips / Roses

   C Tulips / Roses

   D Tulips Roses

Practice 3.G1

III.G *Locate and name points on a number line using whole numbers, fractions [halves and fourths], and decimals [tenths]*

1. Which point best represents 7.4 on the number line?

   A R

   B S

   C T

   D U

2. Which number line shows all the whole numbers that are greater than 1 and less than 4?

   A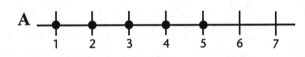

   B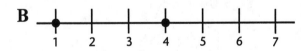

   C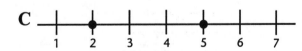

   D

3. On the number line below, the arrow points most closely to—

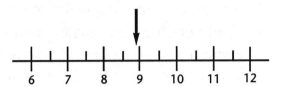

   A 8.2

   B 8.4

   C 8.9

   D 9.1

4. On which number line does the arrow point most closely to $2\frac{2}{3}$?

   A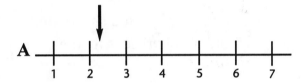

   B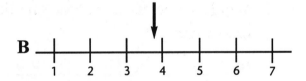

   C

   D

Practice 3.G2

1. What number does point T best represent?

A 11

B 11.2

C 11.5

D 11.7

2. Which number line shows only whole numbers less than 5?

A

B

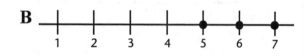

C

D

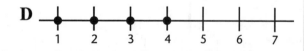

3. Which point best represents $3\frac{1}{3}$ on the number line?

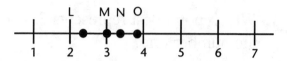

A L

B M

C N

D O

4. On the number line below, the arrow points most closely to—

A 9.6

B 9.4

C 9.2

D 9.0

Practice 3.G3

III.G Locate and name points on a number line using whole numbers, fractions [halves and fourths], and decimals [tenths]

1. What number does point D best represent?

   A 3.1

   B 2.7

   C 2.2

   D 1.9

2. Which number line shows whole numbers more than 2.4 and less than 4.5?

A

B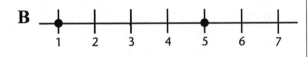

C

D

3. Which point best represents 9.3 on the number line?

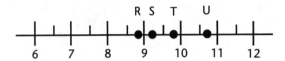

   A R

   B S

   C T

   D U

4. On the number line below, the arrow points most closely to—

   A $6\frac{7}{8}$

   B $6\frac{1}{2}$

   C $6\frac{1}{3}$

   D $7\frac{1}{8}$

Notes

Measurement Concepts

IV. Demonstrate an understanding of measurement concepts, using metric and customary units

    A. Estimate, measure, and compare weight using customary and metric units

    B. Estimate, measure, and compare capacity using customary and metric units

    C. Estimate, measure, and compare length using customary and metric units

    D. Carry out simple unit conversions within a system of measurement

Notes

Objective 4: Pretest

IV.A Estimate, measure, and compare weight using customary and metric units (1-6)

1. About how much does the box of fruit weigh?

   A 10 lb

   B 8 lb

   C 6 lb

   D 4 lb

2. About how much does the average candy bar weigh?

   A 5000 g

   B 500 g

   C 50 g

   D 5 g

3. What is the difference in weight between Box A and Box B?

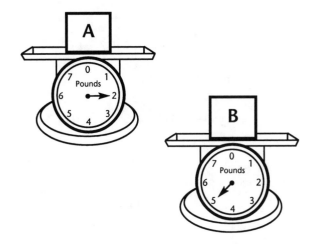

   A 2 lb

   B 3 lb

   C 5 lb

   D 7 lb

4. About how much does the box of candy weigh?

   A 3 kg

   B 2.5 kg

   C 2 kg

   D 1.5 kg

5. About how much does the box of macaroni and cheese weigh?

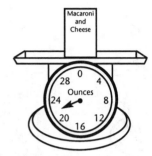

A 18 oz

B 20 oz

C 22 oz

D 24 oz

6. About how much more does Box A weigh than Box B?

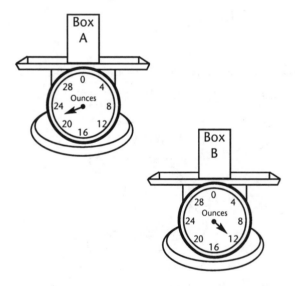

A 20 oz

B 12 oz

C 10 oz

D 8 oz

IV.B Estimate, measure, and compare capacity using customary and metric units (7-15)

7. About how much milk does an average drinking glass hold?

A 1 oz

B 3 oz

C 12 oz

D 32 oz

8. Jerry drank one can of soda. The average can of soda holds about—

A 35 mL

B 350 mL

C 3,500 mL

D 35,000 mL

9. About how much liquid is in the container below?

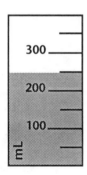

A 200 mL

B 300 mL

C 250 mL

D 275 mL

10. How much more liquid is in container A than in container B?

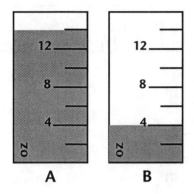

A 8 oz

B 10 oz

C 14 oz

D 18 oz

11. Gretchen made a pitcher of juice. About how much juice does the average pitcher hold?

A 2 c

B 2 pt

C 2 qt

D 2 gal

12. About how much milk might you use in a cookie recipe?

A 2 mL

B 150 mL

C 1,000 mL

D 250 L

13. About how much more liquid is in container B than container A?

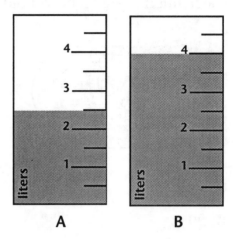

A 6 L

B 2.5 L

C 1.5 L

D 1 L

14. About how much water might you use to fill a small birdbath?

A 3 L

B 30 L

C 3 mL

D 30 mL

15. Janet keeps two angelfish in a small fish bowl. About how much water might the fish bowl hold?

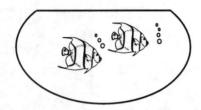

A 5 oz

B 50 oz

C 50 c

D 50 gal

IV.C Estimate, measure, and compare length using customary and metric units (16-22)

16. Mr. Tron bought a new belt. About how long is a man's belt?

A 3 in

B 3 ft

C 3 yd

D 5 yd

17. About how tall is the average door in a house?

A 200 m

B 20 m

C 2 m

D 1 m

18. About how long is this line?

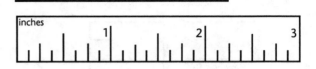

A $2\frac{1}{4}$ in

B $1\frac{3}{4}$ in

C $2\frac{1}{2}$ in

D 2 in

19. How much longer is line A than line B?

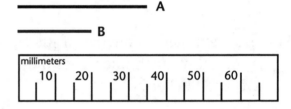

A 25 mm

B 20 mm

C 15 mm

D 10 mm

20. About how long is the average bed?

A 6 yd

B 12 ft

C 1 yd

D 6 ft

21. About how long is this line?

_____

A 600 cm

B 60 cm

C 30 cm

D 6 cm

22. About how long is this line?

_____

A 3 in

B 6 in

C 10 in

D 13 in

IV.D Carry out simple unit conversions within a system of measurement (23-28)

23. A bookstand is $3\frac{1}{2}$ feet tall. How many inches is that?

A 30 in

B 36 in

C 40 in

D 42 in

24. A baseball bat is 1,200 millimeters long. How many meters is that?

A 120 m

B 12 m

C 1.2 m

D 0.12 m

25. A large cooking pot holds 3.8 liters of water. How many milliliters does the pot hold?

A 38 mL

B 380 mL

C 3,800 mL

D 38,000 mL

26. Donna drank 3 pints of milk. How many cups of milk did she drink?

A 12 c

B 6 c

C 4 c

D 2 c

27. Johnny bought $2\frac{1}{2}$ pounds of candy. How many ounces of candy did he buy?

A 24 oz

B 32 oz

C 36 oz

D 40 oz

28. A bag holds 2,400 grams of potatoes. How many kilograms is that?

A 2.4 kg

B 24 kg

C 240 kg

D 2,400 kg

Practice 4.A1

1. About how much does the bag of marbles weigh?

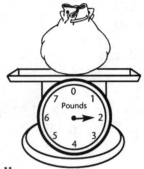

    A 8 lb

    B 6 lb

    C 4 lb

    D 2 lb

2. About how much does a bar of bath soap weigh?

    A 3 g

    B 30 g

    C 300 g

    D 3,000 g

3. About how much might a large dog weigh?

    A 3 kg

    B 30 kg

    C 300 kg

    D 3,000 kg

4. What is the difference in weight between Box A and Box B?

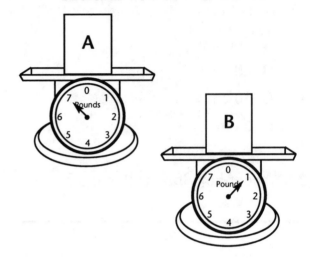

    A 8 lb

    B 7 lb

    C 6 lb

    D 1 lb

5. About how much do the books weigh?

    A 10 lb

    B 5 lb

    C 4 lb

    D 1 lb

Practice 4.A2

IV.A Estimate, measure, and compare weight using customary and metric units

1. About how much do the apples weigh?

   A 10 oz

   B 12 oz

   C 14 oz

   D 16 oz

2. About how much does an average textbook weigh?

   A 1 lb

   B 15 lb

   C 50 lb

   D 500 lb

3. About how much does the box weigh?

   A 2 lb

   B 4 lb

   C 5 lb

   D 7 lb

4. What is the difference in weight between Box A and Box B?

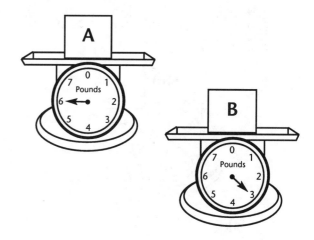

   A 3 lb

   B 6 lb

   C 9 lb

   D 12 lb

5. What is the difference in weight between the two bags?

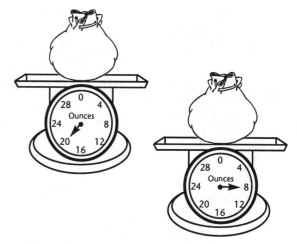

   A 28 oz

   B 20 oz

   C 12 oz

   D 10 oz

Practice 4.A3

1. About how much does the box of fruit weigh?

   A 2 kg

   B 6 kg

   C 8 kg

   D 16 kg

2. About how much does an apple weigh?

   A 4 g

   B 40 g

   C 400 g

   D 4,000 g

3. About how much might a newborn baby weigh?

   A 20 kg

   B 10 kg

   C 4 kg

   D 1 kg

4. How much do both boxes weigh together?

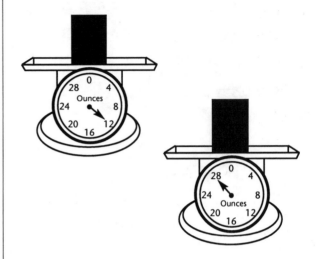

   A 40 oz

   B 30 oz

   C 16 oz

   D 12 oz

5. What is the difference in weight between Box A and Box B?

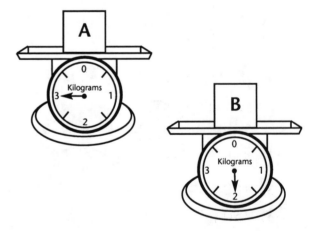

   A 1 kg

   B 2 kg

   C 5 kg

   D 6 kg

Practice 4.B1

IV.B Estimate, measure, and compare capacity using customary and metric units

1. About how much orange juice does an average drinking glass hold?

   A 2 oz

   B 4 oz

   C 12 oz

   D 32 oz

2. Marty and Cindy drank one large jug of milk. The average jug of milk holds about—

   A 1 gal

   B 1 pt

   C 1 c

   D 1 oz

3. About how much liquid is in the container below?

   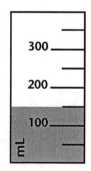

   A 100 mL

   B 150 mL

   C 175 mL

   D 200 mL

4. How much more liquid is in container A than in container B?

   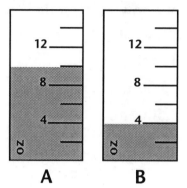

   A 4 oz

   B 6 oz

   C 8 oz

   D 12 oz

5. If you combine containers A and B, how much liquid will you have?

   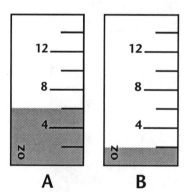

   A 4 oz

   B 5 oz

   C 6 oz

   D 8 oz

Practice 4.B2

IV.B *Estimate, measure, and compare capacity using customary and metric units*

1. About how much soda is in the average can?

   A 35 mL

   B 350 mL

   C 3,500 mL

   D 35,000 mL

2. Anna ate a bowl of soup. About how many ounces of soup did she eat?

   A 16 oz

   B 40 oz

   C 80 oz

   D 800 oz

3. About how much liquid is in the container below?

   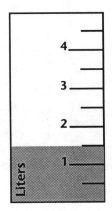

   A 0.75 L

   B 1 L

   C 1.25 L

   D 1.5 L

4. How much more liquid is in container A than in container B?

   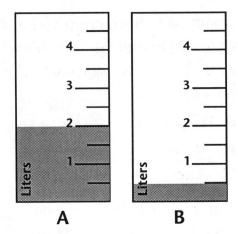

   A 1.5 L

   B 0.75 L

   C 0.5 L

   D 0.25 L

5. If you combine containers A and B, how much liquid will you have?

   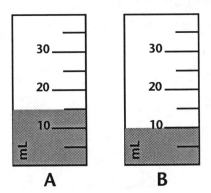

   A 30 mL

   B 25 mL

   C 20 mL

   D 15 mL

Practice 4.B3

IV.B Estimate, measure, and compare capacity using customary and metric units

1. A pitcher holds enough lemonade to fill 6 drinking glasses. About how much lemonade does the pitcher hold?

   A 6,000 oz

   B 700 oz

   C 72 oz

   D 6 oz

2. About how much milk might you add to a bowl of cereal?

   A 1 L

   B 250 mL

   C 50 mL

   D 25 mL

3. Peter filled the kitchen sink with water. About how much water did he use?

   A 3 c

   B 3 pt

   C 3 qt

   D 3 gal

4. How much more liquid is in container A than in container B?

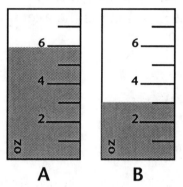

   A 1 oz

   B 2 oz

   C 3 oz

   D 4 oz

5. If you combine the liquid in both containers, how much liquid will you have?

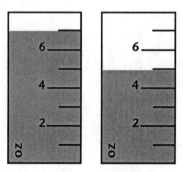

   A 14 oz

   B 12 oz

   C 4 oz

   D 3 oz

Practice 4.C1

IV.C *Estimate, measure, and compare length using customary and metric units*

1. Mrs. Sanders bought a new couch. About how long is the average couch?

   A 7 ft

   B 1 yd

   C 17 ft

   D 7 yd

2. Clara wants to know if a computer monitor will fit on her desk. About how wide is the average computer monitor?

   A $1\frac{1}{2}$ yd

   B $3\frac{1}{2}$ ft

   C $1\frac{1}{2}$ ft

   D 1 yd

3. About how long is this line?

   ▬

   A 1 mm

   B 10 mm

   C 5 cm

   D 50 mm

4. How much longer is line A than line B?

   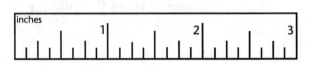

   A 1 in

   B $1\frac{1}{4}$ in

   C $\frac{1}{4}$ in

   D $\frac{3}{4}$ in

5. About how long is this line?

   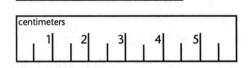

   A 40 cm

   B 0.45 cm

   C 4.5 cm

   D 450 cm

Practice 4.C2

IV.C Estimate, measure, and compare length using customary and metric units

1. About how long is the average textbook?

   A 12 cm

   B 12 ft

   C 100 mm

   D 12 in

2. Monica needs a tablecloth for her rectangular kitchen table. About how long should the tablecloth be?

   A 6 ft

   B 6 yd

   C 2 yd

   D 18 ft

3. About how long is this line?

   A 5 mm

   B 50 mm

   C 50 cm

   D 500 cm

4. How much longer is line A than line B?

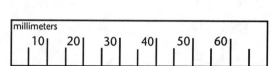

   A 20 mm

   B 25 mm

   C 30 mm

   D 35 mm

5. About how long is this line?

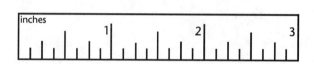

   A $2\frac{1}{4}$ in

   B $2\frac{1}{2}$ in

   C $2\frac{3}{4}$ in

   D 3 in

Practice 4.C3

IV.C Estimate, measure, and compare length using customary and metric units

1. About how wide is the average doorway?

   A 8 yd

   B 18 ft

   C 1 yd

   D 8 ft

2. Mr. Jordan wants to know if his new refrigerator will fit along a wall in his kitchen. About how tall is the average refrigerator?

   A 16 yd

   B 16 ft

   C 6 yd

   D 6 ft

3. About how long is this line?

   A $\frac{3}{4}$ in

   B $1\frac{1}{2}$ in

   C $1\frac{1}{4}$ in

   D $2\frac{1}{4}$ in

4. About how long would line A and line B be together?

   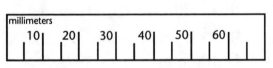

   A 10 mm

   B 15 mm

   C 35 mm

   D 55 mm

5. About how long is this line?

   A 60 cm

   B 6 cm

   C 6,000 mm

   D 6 mm

Practice 4.D1

*IV.D Carry out simple unit conversions within a
system of measurement*

1. Cecile is $5\frac{1}{2}$ feet tall. How many inches tall is she?

   A 55 in

   B 60 in

   C 66 in

   D 70 in

2. Marcus used 2 quarts of oil in his lawn mower. How many pints of oil did he use?

   A 8 pt

   B 6 pt

   C 5 pt

   D 4 pt

3. A bowl holds 2,500 milliliters of water. How many liters does the bowl hold?

   A 2.5 L

   B 10 L

   C 20 L

   D 25 L

4. At the school walk-a-thon, Darren walked 2.3 km. How many meters did he walk?

   A 0.23 m

   B 23 m

   C 230 m

   D 2,300 m

5. A pencil is 0.16 meters long. How many centimeters is that?

   A 160 cm

   B 16 cm

   C 10 cm

   D 1.6 cm

6. Cherisse bought 4 yards of ribbon for a craft project. How many feet of ribbon did she buy?

   A 16 ft

   B 14 ft

   C 12 ft

   D 8 ft

7. The distance around a table is 8 feet. How many inches is that?

   A 96 in

   B 84 in

   C 80 in

   D 16 in

8. Natalie drank 6 cups of milk in one day. How many pints of milk did she drink?

   A 24 pt

   B 12 pt

   C 8 pt

   D 3 pt

Practice 4.D2

1. Nate is 54 inches tall. How many feet tall is he?

   A 6 ft

   B $5\frac{1}{2}$ ft

   C 5 ft

   D $4\frac{1}{2}$ ft

2. The **perimeter** of a square is 480 centimeters. How many meters is that?

   A 0.48 m

   B 4.8 m

   C 48 m

   D 480 m

3. A bag of flour weighs 2.3 kilograms. What is the bag's weight in grams?

   A 2.3 g

   B 23 g

   C 230 g

   D 2,300 g

4. Dennis has 12 feet of rope. How many yards of rope does he have?

   A 2 yd

   B 3 yd

   C 4 yd

   D 6 yd

5. Tony's dad is 180 centimeters tall. How many meters tall is he?

   A 0.18 m

   B 1.8 m

   C 18 m

   D 1,800 m

6. A fish bowl holds 1,900 milliliters of water. How many liters of water does the bowl hold?

   A 1.9 L

   B 19 L

   C 190 L

   D 19,000 L

7. To make a bracelet, Ginny needs 110 millimeters of string. How many centimeters of string does she need?

   A 0.11 cm

   B 1.1 cm

   C 11 cm

   D 1,100 cm

8. Gary took 4 quarts of water on a camping trip. How many gallons of water is that?

   A 1 gal

   B 2 gal

   C 4 gal

   D 16 gal

Practice 4.D3

1. A truck weighs $2\frac{1}{2}$ tons. How many pounds does the truck weigh?

   A 250 lb

   B 1,500 lb

   C 3,000 lb

   D 5,000 lb

2. Joey used 4 pounds of potting soil in a flower box. How many ounces of soil did he use?

   A 64 oz

   B 48 oz

   C 32 oz

   D 24 oz

3. For a science experiment, Jamal used 200 milliliters of alcohol. How many liters of alcohol did he use?

   A 20 L

   B 2 L

   C 0.2 L

   D 0.02 L

4. Sarita bought 3,000 centimeters of ribbon. How many meters of ribbon did she buy?

   A 0.3 m

   B 3 m

   C 30 m

   D 300 m

5. Devonia wants to make a belt. She needs $1\frac{1}{2}$ feet of leather. How many inches is that?

   A 12 in

   B 18 in

   C 24 in

   D 36 in

6. The **perimeter** of Mrs. Rashad's classroom is 120 feet. What is the perimeter of the room in yards?

   A 6 yd

   B 10 yd

   C 30 yd

   D 40 yd

7. On a hiking trip, Joy walked 3,100 meters. How many kilometers did she walk?

   A 0.31 km

   B 3.1 km

   C 31 km

   D 310 km

8. A box of cookies weighs 1,900 grams. What is the box's weight in kilograms?

   A 190 kg

   B 19 kg

   C 1.9 kg

   D 0.19 kg

Notes

Probability and Statistics

V. Demonstrate an understanding of probability and statistics

A. List possible outcomes of a probability experiment (e.g., tossing a coin)

B. Make predictions based on a sampling

C. Determine the mean (average), median, and mode from collected data

D. Interpret bar graphs, tables, and charts

Notes

Objective 5: Pretest

V.A List possible outcomes of a probability experiment [e.g., tossing a coin] (1-5)

1. A bag holds 5 marbles. They are the same size and shape, but have different colors.

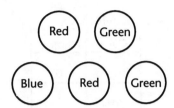

   Which is a possible outcome if 3 marbles are selected from the bag all at one time?

   A Red   Blue   Blue

   B Green   Red   Blue

   C Blue   Blue   Green

   D Red   Red   Red

2. There were 6 pieces of candy in a box. The candies looked the same, but they had different flavors.

   Jody took 3 pieces of candy from the box all at one time. Which is a possible combination of candy Jody could have taken?

   A 2 vanilla and 1 chocolate

   B 3 cherry

   C 2 toffee and 1 vanilla

   D 1 toffee and 2 cherry

3. Damon is playing cards with his sister. He has 7 cards in his hand.

   If he plays 3 cards on his next turn, which combination of cards could he play?

   A J Q J

   B A A A

   C A Q Q

   D K K J

4. Donna is playing a game on the board shown below. If she throws a marker 15 times, which number will the marker probably land on the **greatest** number of times?

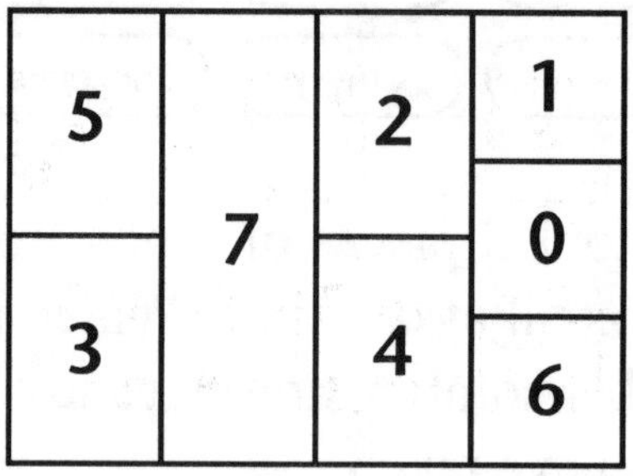

			1
5	7	2	0
3		4	6

A 1

B 3

C 5

D 7

5. Mrs. Garza keeps small erasers in a prize box. The erasers are the same size and shape, but have different colors.

Erasers

Color	Number
Red	4
Blue	6
Green	2
Yellow	3

If Joyce takes an eraser without looking in the box, what color will the eraser **most** likely be?

A Green

B Blue

C Red

D Yellow

V.B Make predictions based on a sampling (6-9)

6. A box holds 10 marbles. Denice removes 5 marbles from the box, all at one time. The chart shows which marbles she took from the box.

Marbles

Color	Number
Striped	0
Clear	1
Black	3
Red	1

Which of these is **most** likely to be true?

A There are more black marbles in the box than any other kind.

B There is only 1 red marble in the box.

C There are more clear marbles in the box than red marbles.

D There are more striped marbles in the box than any other kind.

7. Martin spun a game wheel 10 times. The chart shows the results of his spins.

Spins

Symbol	Number
❤	1
✪	6
✳	2
✿	1

Which game wheel was he **most** likely using?

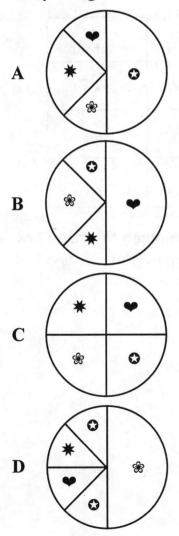

8. There are 50 game cards in a box. Each card has from 1 to 6 spots on it. Leann pulled 10 cards from the box without looking. These are the cards she pulled.

Which of these is **most** likely to be true?

A Most cards in the box have 1 spot.

B Most cards in the box have 2 spots.

C Most cards in the box have 3 spots.

D Most cards in the box have 4 spots.

9. Ileana played a beanbag game with her friend. The chart shows the number of times her beanbag landed on each color shown on the game board.

Beanbag Tosses

Color	Number
Pink (P)	2
Green (G)	2
Red (R)	5
Blue (B)	1

Which game board was Ileana **most** likely using?

A

P	G	R	B	B
P	G	R	B	B

B

P	P	P	P	B
P	R	R	B	G

C

P	G	R	R	R
P	G	R	R	B

D

P	G	R	B	G
P	G	R	B	G

V.C Determine the mean [average], median, and mode from collected data (10-15)

10. In 3 games, Jamal's team scored 16, 12, and 8 points. What is the team's **mean** (average) score for those 3 games?

A 36

B 18

C 12

D 10

11. The chart shows the low temperatures for 3 days during December.

Date	Low Temperature
12/3	32°F
12/4	36°F
12/5	34°F

What is the **mean** (average) low temperature for those 3 days?

A 34°F

B 36°F

C 51°F

D 102°F

12. Vicki bowled 5 games. These were her scores: 101, 104, 108, 110, and 112. What was Vicki's **median** score?

A 112

B 108

C 107

D 100

13. There are 5 fourth-grade classes in the school. The chart shows the number of children in each class.

Class	Students
A	19
B	21
C	24
D	25
E	26

What is the **median** number of students in the fourth-grade classes?

A 19

B 23

C 24

D 26

14. Clark earned these scores on his spelling tests: 95, 90, 95, 85, 95, 100, and 80. Which number is the **mode** of Clark's spelling test scores?

A 100

B 95

C 90

D 85

15. The chart shows the high temperature for one week in May.

Day	High Temperature
Sunday	78°F
Monday	76°F
Tuesday	76°F
Wednesday	70°F
Thursday	75°F
Friday	76°F
Saturday	73°F

Which one is the **mode** of these high temperatures?

A 70°F

B 73°F

C 75°F

D 76°F

V.D Interpret bar graphs, tables, and charts (16-19)

Use the bar graph to answer questions 16 and 17.

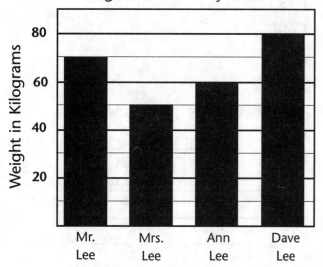

Weight of Lee Family Members

16. How much more does Dave Lee weigh than Mr. Lee?

    A 5 kg

    B 10 kg

    C 15 kg

    D 20 kg

17. How much does Mrs. Lee weigh?

    A 58 kg

    B 52 kg

    C 50 kg

    D 45 kg

Use the table to answer questions 18-20.

Fourth-Grade Absences

Date	Boys	Girls
Jan. 3	12	7
Jan. 4	8	3
Jan. 5	15	8
Jan. 6	6	5
Jan. 7	9	16

18. On what date were the **greatest** number of boys absent?

    A Jan. 7

    B Jan. 5

    C Jan. 4

    D Jan. 3

19. On what date were more girls absent than boys?

    A Jan. 3

    B Jan. 4

    C Jan. 6

    D Jan. 7

20. How many fourth-grade students were absent on January 4?

    A 3

    B 5

    C 8

    D 11

Practice 5.A1

1. Peter had 6 Bouncy Balls. They were the same size and shape, but had different designs.

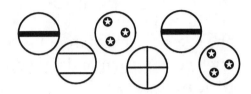

Peter gave away 3 Bouncy Balls. Which is a possible combination of the Bouncy Balls he gave away?

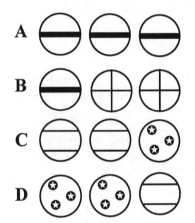

2. There are 3 cherry gumdrops, 2 lime gumdrops, and 5 lemon gumdrops in a bag. Which possible combination of gumdrops could Sammy take from the bag?

   A 2 cherry, 3 lime

   B 2 cherry, 4 lemon

   C 4 cherry, 2 lime

   D 3 lime, 3 lemon

3. Darian is playing cards with his mother. He has 8 cards in his hand.

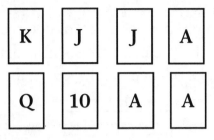

If Darian plays 4 cards on his next turn, which combination of cards could he play?

   A K J A A

   B A A 10 10

   C A Q Q K

   D J Q J Q

4. Nikki played a game with the spinner shown below. She spun 15 times. Which picture did she probably land on **most** often?

   A ✳

   B ❁

   C ☎

   D ❤

Practice 5.A2

1. Lauren bought a box of colored chalk. The chart shows how many sticks of each color were in the box.

Chalk

Color	Number
Orange	2
Blue	5
White	8
Yellow	5

If Lauren chooses one stick of chalk without looking in the box, what color will it **most** likely be?

A Blue

B White

C Orange

D Yellow

2. Tina has 4 red bracelets, 3 green bracelets, 2 gold bracelets, and 1 blue bracelet. Which combination of bracelets could Tina wear?

A 2 red, 2 blue, 2 gold

B 1 red, 2 gold, 3 blue

C 1 green, 1 gold, 1 blue

D 3 green, 3 gold, 3 red

3. Brendan has 9 pennies in his pocket.

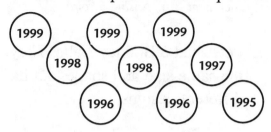

If he takes 4 pennies from his pocket, which combination of dates could be on the pennies?

A 1999, 1999, 1999, 1999

B 1999, 1999, 1997, 1997

C 1997, 1997, 1996, 1995

D 1998, 1998, 1997, 1995

4. Janna is playing a game on the board shown below. If she throws a marker 20 times, which number will it probably land on the **greatest** number of times?

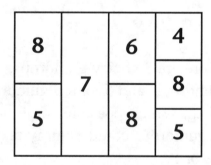

A 5

B 6

C 7

D 8

Practice 5.A3

V.A List possible outcomes of a probability experiment (e.g., tossing a coin)

1. Mrs. Avery made 20 prize tickets for a class game. The chart shows how many tickets she made for each prize.

Tickets

Prize	Number
Candy	6
Pencil	2
Ruler	8
Eraser	14

If Tim chooses a prize ticket without looking, what prize is he **least** likely to choose?

A Candy

B Pencil

C Ruler

D Eraser

2. A box contains 2 gold beads, 4 silver beads, 1 copper bead, and 3 bronze beads. Which combination of beads could you take from the box?

A 2 gold, 2 silver, 2 bronze

B 4 gold, 2 copper, 2 bronze

C 2 silver, 1 copper, 4 bronze

D 3 copper, 2 gold, 1 silver

3. Brandy has 7 game tokens. The tokens are the same size and shape, but have different numbers of points.

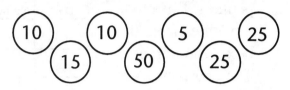

If she uses 3 tokens during the game, which combination of tokens could she play?

A 5, 5, 10

B 10, 10, 10

C 10, 10, 50

D 25, 50, 50

4. Richard plays a word game with the spinner shown below. If he spins 20 times, which word will he probably land on **most** often?

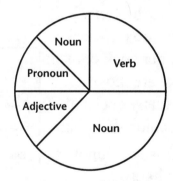

A Noun

B Verb

C Adjective

D Pronoun

Practice 5.B1

1. A box held 15 pennies. Elisa took 8 pennies from the box at one time. The chart shows the dates on the pennies she took from the box.

Pennies

Date	Number
1996	2
1997	1
1998	5
1999	0

Which of these is **most** likely true?

A There were more 1999 pennies in the box than any other date.

B There were the same number of 1998 and 1999 pennies in the box.

C There were more 1996 pennies than 1997 pennies in the box.

D There were more 1997 pennies in the box than any other date.

2. Without looking, Karen took 10 socks from her sock drawer. The chart shows how many socks of each color she took.

Socks

Color	Number
Red	1
Green	1
Brown	3
Black	5

Karen's sock drawer **most** likely—

A had more black than red socks

B did not have green socks

C had the same number of green and black socks

D had more red than green socks

3. Thomas spun a game wheel 12 times. The chart shows the results of his spins.

Spins

Symbol	Number
④	2
❺	3
❶	1
⑦	6

Which game wheel was he **most** likely using?

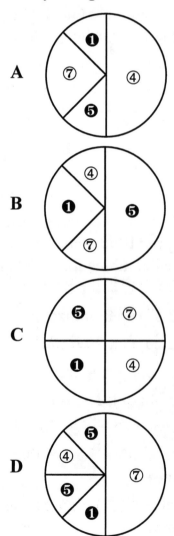

Practice 5.B2

V.B Make predictions based on a sampling

1. A box held 20 tickets. A name of a prize was printed on each ticket. Jeff took 10 tickets from the box all at one time. The chart shows the prizes for the tickets he took from the box. Which of these is **most** likely true?

Tickets	
Prize	Number
Whistle	4
Yo-yo	2
Pen Set	3
Candy	1

   A Most tickets in the box were for a yo-yo.

   B There were more tickets in the box for candy than for a pen set.

   C There was a greater chance of winning a whistle than a yo-yo.

   D There were 10 tickets for candy in the box.

2. A bag held 40 lollipops. Sherry took 8 lollipops from the bag without looking. The chart shows how many of each flavor she took. The bag **most** likely—

Lollipops	
Flavor	Number
Cherry	4
Lime	1
Lemon	2
Orange	1

   A had 20 lime lollipops

   B had exactly 4 cherry lollipops

   C had more lemon lollipops than cherry ones

   D had fewer orange lollipops than cherry ones

3. Inez played a beanbag game with her friend. The chart shows the number of times her beanbag landed on each color shown on the game board.

Beanbag Toss	
Color	Number
Yellow (Y)	1
Green (G)	6
Purple (P)	3
Blue (B)	2

Which game board was she **most** likely using?

A

B	Y	B	Y	B
P	B	G	B	P

B

G	Y	B	G	P
G	P	G	B	G

C

G	G	Y	P	Y
G	B	Y	B	Y

D

G	Y	B	Y	G
B	P	Y	B	P

Practice 5.B3

1. A bag held 50 glass beads. The beads were the same size and shape, but different colors. Without looking, Pearl took 15 beads from the bag. The chart shows how many of each color she took. Which is **most** likely true?

   Beads

Color	Number
White	4
Pink	7
Green	3
Blue	1

   A There were only 7 pink beads in the bag.

   B There were more blue beads than white beads in the bag.

   C There were fewer pink beads than green beads in the bag.

   D There were fewer blue beads than any other color in the bag.

2. There were 100 marbles in a box. Without looking, Nick took 10 marbles from the box. The chart shows how many of each design he took.

   Marbles

Design	Number
Clear	5
Striped	1
Stars	2
Swirled	2

   The box **most** likely had—

   A 50 swirled marbles

   B more clear marbles than striped ones

   C the same number of striped and clear marbles

   D 50 marbles with stars

3. Jake played a dart game with his brother. The chart shows how many times Jake's dart hit each number on the dart board.

   Dart   Throws

Number	Hits
25	1
15	2
10	3
5	6

   Which dart board was Jake **most** likely using?

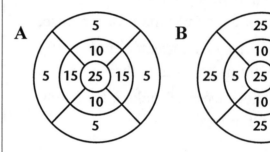

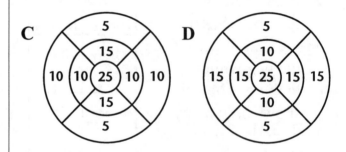

Practice 5.C1

V.C Determine the mean [average], median, and mode from collected data

1. Denzel earned the following scores on his math tests: 92, 96, and 91. What was his **mean** (average) score on the three tests?

A 96

B 91

C 94

D 93

2. The chart shows the high temperatures for 3 days during March.

Date	High Temperature
3/10	57°F
3/11	51°F
3/12	48°F

What is the **mean** (average) high temperature for these 3 days?

A 52°F

B 51°F

C 54°F

D 56°F

3. A basketball team played five games. These were their scores: 46, 49, 45, 50, and 50. What was the team's **median** score?

A 45

B 48

C 49

D 50

4. The school nurse recorded the heights of several students. These were their heights: 56 in, 58 in, 52 in, 56 in, 52 in, 56 in, 57 in, and 53 in. Which is the **mode** of the students' heights?

A 56 in

B 55 in

C 53 in

D 52 in

5. Mr. Flynn bowled 7 games. These were his scores: 150, 156, 189, 200, 175, 162, and 193. What is the **median** of those scores?

A 162

B 175

C 189

D 200

Practice 5.C2

V.C Determine the mean [average], median, and mode from colledted data

1. A book company received 132 orders in May, 114 orders in June, and 126 orders in July. What is the **mean** (average) number of orders the company received each month?

A 114

B 124

C 126

D 134

2. The chart shows the scores earned by 4 teams in the school Think-a-Thon.

Team	Score
A	540
B	625
C	575
D	560

What is the **mean** (average) score earned by these 4 teams?

A 625

B 585

C 575

D 565

3. Christopher's last 5 golf scores were 72, 80, 76, 68, and 74. What is the **median** of those golf scores?

A 68

B 72

C 74

D 76

4. Judy times how long it takes her to walk from her home to her best friend's house. These are her most recent times: 20 min, 18 min, 21 min, 23 min, 18 min, 22 min, and 18 min. Which is the **mode** of those times?

A 18 min

B 20 min

C 21 min

D 23 min

5. Bobby earned the following scores on his math quizzes: 79, 82, 85, 79, 89, 92, and 75. What is the **median** of Bobby's score?

A 92

B 89

C 82

D 79

Practice 5.C3

V.C Determi[illegible]e], median, and mod[illegible]

1. [illegible]223 miles [illegible]Tuesday, an[illegible]day. What is th[illegible]umber of miles th[illegible]ch day?

   A 816 mi

   B 205 mi

   C 272 mi

   D 262 mi

2. The chart shows the number of students in 4 fifth-grade classes.

Class	Students
A	26
B	22
C	24
D	28

   What is the **mean** (average) number of students in the fifth-grade classes?

   A 25

   B 23

   C 27

   D 22

3. For babysitting, Carla earned $80 in May, $85 in June, $60 in July, $80 in August, and $70 in September. What is the **median** amount Carla earned during those 5 months?

   A $60

   B $75

   C $80

   D $85

4. Aaron practices playing the piano every day. These are the number of minutes he practiced in one week: 45 min, 65 min, 50 min, 35 min, 75 min, 70 min, and 45 min. What is the **median** amount of time Aaron practiced?

   A 35 min

   B 65 min

   C 55 min

   D 50 min

5. The high temperatures in Dallas for one week were 88°F, 83°F, 94°F, 88°F, 97°F, 92°F, and 88°F. Which is the **mode** of those high temperatures?

   A 90°F

   B 88°F

   C 97°F

   D 83°F

Practice 5.D1

V.D Interpret bar graphs, tables, and charts

The bar graph shows how many birds Carl saw at his feeder in one week. Use the graph to answer questions 1 and 2.

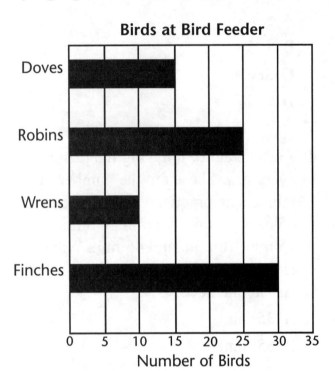

Birds at Bird Feeder

1. How many wrens did Carl see at the bird feeder?

   A 2

   B 5

   C 10

   D 15

2. How many more finches than doves went to the bird feeder?

   A 30

   B 25

   C 20

   D 15

The chart shows how many students go to special classes each day. Use the chart to answer questions 3-5.

Special Class Schedule

	Art	Music	PE
Mon.	65	80	125
Tues.	45	90	135
Wed.	85	45	140
Thurs.	75	75	120
Fri.	65	75	130

3. How many students go to art on Friday?

   A 130

   B 85

   C 75

   D 65

4. On Monday, how many more students go to music than to art?

   A 15

   B 45

   C 65

   D 80

5. On which day do the fewest number of students go to PE?

   A Friday

   B Thursday

   C Wednesday

   D Tuesday

Practice 5.D2

V.D Interpret bar graphs, tables, and charts

The line graph shows the level of water in a well during 7 hours. Use the graph to answer questions 1-6.

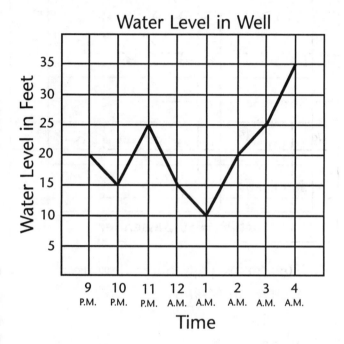

1. How much water was in the well at 12 A.M.?

   A 3 ft

   B 10 ft

   C 15 ft

   D 20 ft

2. How much more water was in the well at 2 A.M. than at 1 A.M.?

   A 30 ft

   B 20 ft

   C 15 ft

   D 10 ft

3. How much less water was in the well at 10 P.M. than at 4 A.M.?

   A 20 ft

   B 25 ft

   C 30 ft

   D 50 ft

4. The **greatest** drop in the water level happened between—

   A 1 A.M. and 2 A.M.

   B 3 A.M. and 4 A.M.

   C 11 P.M. and 12 A.M.

   D 9 P.M. and 10 P.M.

5. How much water was in the well at 3 A.M.?

   A 25 ft

   B 20 ft

   C 15 ft

   D 5 ft

6. The well had the **lowest** level of water at—

   A 9 P.M.

   B 10 P.M.

   C 12 A.M.

   D 1 A.M.

Practice 5.D3

V.D Interpret bar graphs, tables, and charts

The chart shows the lengths of several rivers in Texas. Use the chart to answer questions 1 and 2.

Rivers in Texas

River	Length
Brazos	840 miles
Colorado	600 miles
Neches	416 miles
San Antonio	180 miles
Trinity	550 miles

1. How much longer is the Brazos River than the San Antonio River?

   A 1,020 mi

   B 760 mi

   C 740 mi

   D 660 mi

2. Which two rivers have a combined length that is less than the length of the Colorado River?

   A Brazos and Neches

   B Neches and Trinity

   C San Antonio and Neches

   D San Antonio and Trinity

The graph shows how many passengers flew from Denver to other cities. Use the graph to answers questions 3-5.

Passengers on Planes Leaving Denver

New York	✈ ✈ ✈ ✈ ✈
Dallas	✈ ✈ ✈ ✈
Boston	✈ ✈
Miami	✈ ✈ ✈ ✈ ✈ ✈
San Diego	✈ ✈ ✈

Each ✈ = 50 passengers

3. How many passengers flew from Denver to Miami?

   A 150

   B 300

   C 325

   D 400

4. How many more passengers flew to Dallas than to Boston?

   A 2

   B 50

   C 100

   D 300

5. 150 passengers flew to—

   A New York

   B Dallas

   C Miami

   D San Diego

Appendix

- **Answer Key**
- **Answer Sheets**

Answer Key: Number Concepts

Objective 1 Pretest (p. 11)

1. C	2. B	3. D	4. C	5. B
6. C	7. A	8. A	9. D	10. C
11. D	12. B	13. A	14. C	15. B
16. D	17. B	18. D	19. C	20. C
21. B	22. C	23. A	24. D	25. C
26. D	27. C	28. A	29. D	30. C
31. B	32. C	33. C	34. A	35. D
36. C	37. D	38. B	39. C	40. A

Practice 1.A1 (p. 18)

1. C	2. C	3. A	4. B	5. D
6. A	7. D			

Practice 1.A2 (p. 19)

1. C	2. D	3. A	4. B	5. D
6. D	7. C			

Practice 1.A3 (p. 20)

1. B	2. D	3. A	4. D	5. C
6. C	7. D			

Practice 1.B1 (p. 21)

1. D	2. C	3. A	4. D	5. B
6. C	7. A	8. B		

Practice 1.B2 (p. 22)

1. C	2. B	3. A	4. D	5. C
6. B	7. A	8. B		

Practice 1.B3 (p. 23)

1. D	2. C	3. B	4. A	5. C
6. D	7. A	8. B		

Practice I.C1 (p. 24)

1. C	2. D	3. B	4. C	5. A
6. C	7. D			

Practice 1.C2 (p. 25)

1. A	2. B	3. D	4. C	5. C
6. A	7. D			

Practice 1.C3 (p. 26)

1. C	2. D	3. B	4. D	5. B
6. C	7. D			

Practice 1.D1 (p. 27)

1. C	2. A	3. D	4. A	5. B
6. D				

Practice 1.D2 (p. 28)

1. B	2. D	3. B	4. D	5. C
6. C				

Practice 1.D3 (p. 29)

1. A	2. C	3. C	4. B	5. B
6. C				

Practice 1.E1 (p. 30)

1. A	2. D	3. C	4. C	5. D

Practice 1.E2 (p. 31)

1. C	2. B	3. C	4. C	5. D
6. B				

Practice 1.E3 (p. 32)

1. D	2. B	3. C	4. A	5. B
6. D				

Answer Key: Mathematical Relations, Functions, & Algebraic Concepts

Objective 2 Pretest (p. 35)

1. D	2. B	3. B	4. C	5. A
6. B	7. C	8. D	9. B	10. D
11. C	12. B	13. A	14. C	15. C
16. B	17. B	18. C		

Practice 2. A1 (p. 39)

1. B	2. C	3. D	4. C	5. B

Practice 2. A2 (p. 40)

1. C	2. D	3. B	4. D	5. B

Practice 2. A3 (p. 41)

1. D	2. C	3. D	4. D	5. A

Practice 2. B1 (p. 42)

1. A	2. C	3. C	4. A	5. D
6. B				

Practice 2. B2 (p. 43)

1. D	2. A	3. C	4. A	5. B
6. B				

Practice 2. B3 (p. 44)

1. D	2. C	3. A	4. C	5. B
6. C				

Practice 2. C1 (p. 45)
1. B 2. D 3. D 4. C

Practice 2. C2 (p. 46)
1. C 2. D 3. B 4. A

Practice 2. C3 (p. 47)
1. C 2. D 3. A 4. B

Answer Key: Geometric Properties/Relationships

Objective 3 Pretest (p. 51)
1. C 2. D 3. B 4. A 5. A
6. D 7. C 8. D 9. A 10. B
11. D 12. C 13. D 14. C 15. A
16. B 17. D 18. C 19. C 20. D
21. C 22. B 23. A 24. D 25. A
26. D 27. C 28. D 29. C 30. B
31. D 32. C 33. D 34. A 35. B
36. C 37. C 38. B

Practice 3. A1 (p. 61)
1. A 2. B 3. B 4. D 5. B

Practice 3. A2 (p. 62)
1. B 2. A 3. A 4. D 5. C

Practice 3. A3 (p. 63)
1. A 2. C 3. C 4. D 5. A

Practice 3. B1 (p. 64)
1. A 2. C 3. C 4. B 5. D

Practice 3. B2 (p. 65)
1. C 2. A 3. D 4. C 5. C

Practice 3. B3 (p. 66)
1. D 2. C 3. C 4. D 5. A

Practice 3. C1 (p. 67)
1. D 2. C 3. B 4. C 5. A
6. B

Practice 3. C2 (p. 68)
1. C 2. D 3. A 4. C 5. B
6. D

Practice 3. C3 (p. 69)
1. D 2. C 3. B 4. A 5. C
6. A

Practice 3. D1 (p. 70)
1. D 2. B 3. C 4. D 5. B

Practice 3. D2 (p. 71)
1. D 2. D 3. B 4. B 5. B

Practice 3. D3 (p. 72)
1. B 2. A 3. B 4. D 5. B

Practice 3. E1 (p. 73)
1. C 2. D 3. A

Practice 3. E2 (p. 74)
1. B 2. D 3. C

Practice 3. E3 (p. 75)
1. D 2. B 3. C

Practice 3. F1 (p. 76)
1. C 2. B 3. A 4. D

Practice 3. F2 (p. 77)
1. D 2. B 3. C 4. C

Practice 3. F3 (p. 78)
1. D 2. A 3. B 4. D

Practice 3. G1 (p. 79)
1. B 2. D 3. C 4. C

Practice 3. G2 (p. 80)
1. B 2. D 3. C 4. A

Practice 3. G3 (p. 81)
1. C 2. C 3. B 4. A

Answer Key: Measurement Concepts

Objective 4 Pretest (p. 85)
1. C 2. C 3. B 4. D 5. C
6. C 7. C 8. B 9. C 10. B
11. C 12. B 13. C 14. A 15. B
16. B 17. C 18. A 19. C 20. D
21. D 22. A 23. D 24. C 25. C
26. B 27. D 28. A

Practice 4. A1 (p. 90)
1. D   2. C   3. B   4. C   5. B

Practice 4. A2 (p. 91)
1. C   2. A   3. B   4. A   5. C

Practice 4. A3 (p. 92)
1. A   2. C   3. C   4. A   5. A

Practice 4. B1 (p. 93)
1. C   2. A   3. B   4. B   5. D

Practice 4. B2 (p. 94)
1. B   2. A   3. D   4. A   5. B

Practice 4. B3 (p. 95)
1. C   2. B   3. D   4. C   5. B

Practice 4. C1 (p. 96)
1. A   2. C   3. B   4. D   5. C

Practice 4. C2 (p. 97)
1. D   2. A   3. B   4. D   5. C

Practice 4. C3 (p. 98)
1. C   2. D   3. C   4. D   5. B

Practice 4. D1 (p. 99)
1. C   2. D   3. A   4. D   5. B
6. C   7. A   8. D

Practice 4. D2 (p. 100)
1. D   2. B   3. D   4. C   5. B
6. A   7. C   8. A

Practice 4. D3 (p. 101)
1. D   2. A   3. C   4. C   5. B
6. D   7. B   8. C

Practice 5. A1 (p. 111)
1. D   2. B   3. A   4. C

Practice 5. A2 (p. 112)
1. B   2. C   3. D   4. D

Practice 5. A3 (p. 113)
1. B   2. A   3. C   4. A

Practice 5. B1 (p. 114)
1. C   2. A   3. D

Practice 5. B2 (p. 115)
1. C   2. D   3. B

Practice 5. B3 (p. 116)
1. D   2. B   3. A

Practice 5. C1 (p. 117)
1. D   2. A   3. C   4. A   5. B

Practice 5. C2 (p. 118)
1. B   2. C   3. C   4. A   5. C

Practice 5. C3 (p. 119)
1. C   2. A   3. C   4. D   5. B

Practice 5. D1 (p. 120)
1. C   2. D   3. D   4. A   5. B

Practice 5. D2 (p. 121)
1. C   2. D   3. A   4. C   5. A
6. D

Practice 5. D3 (p. 122)
1. D   2. C   3. B   4. C   5. D

Answer Key: Probability and Statistics

Objective 5 Pretest (p. 105)
1. B   2. D   3. A   4. D   5. B
6. A   7. A   8. D   9. C   10. C
11. A   12. B   13. C   14. B   15. D
16. B   17. C   18. B   19. D   20. D

Name _____   **Date** _____

Objective # _____ Pretest

Pretest Answer Sheet

1. Ⓐ Ⓑ Ⓒ Ⓓ   11. Ⓐ Ⓑ Ⓒ Ⓓ   21. Ⓐ Ⓑ Ⓒ Ⓓ   31. Ⓐ Ⓑ Ⓒ Ⓓ

2. Ⓐ Ⓑ Ⓒ Ⓓ   12. Ⓐ Ⓑ Ⓒ Ⓓ   22. Ⓐ Ⓑ Ⓒ Ⓓ   32. Ⓐ Ⓑ Ⓒ Ⓓ

3. Ⓐ Ⓑ Ⓒ Ⓓ   13. Ⓐ Ⓑ Ⓒ Ⓓ   23. Ⓐ Ⓑ Ⓒ Ⓓ   33. Ⓐ Ⓑ Ⓒ Ⓓ

4. Ⓐ Ⓑ Ⓒ Ⓓ   14. Ⓐ Ⓑ Ⓒ Ⓓ   24. Ⓐ Ⓑ Ⓒ Ⓓ   34. Ⓐ Ⓑ Ⓒ Ⓓ

5. Ⓐ Ⓑ Ⓒ Ⓓ   15. Ⓐ Ⓑ Ⓒ Ⓓ   25. Ⓐ Ⓑ Ⓒ Ⓓ   35. Ⓐ Ⓑ Ⓒ Ⓓ

6. Ⓐ Ⓑ Ⓒ Ⓓ   16. Ⓐ Ⓑ Ⓒ Ⓓ   26. Ⓐ Ⓑ Ⓒ Ⓓ   36. Ⓐ Ⓑ Ⓒ Ⓓ

7. Ⓐ Ⓑ Ⓒ Ⓓ   17. Ⓐ Ⓑ Ⓒ Ⓓ   27. Ⓐ Ⓑ Ⓒ Ⓓ   37. Ⓐ Ⓑ Ⓒ Ⓓ

8. Ⓐ Ⓑ Ⓒ Ⓓ   18. Ⓐ Ⓑ Ⓒ Ⓓ   28. Ⓐ Ⓑ Ⓒ Ⓓ   38. Ⓐ Ⓑ Ⓒ Ⓓ

9. Ⓐ Ⓑ Ⓒ Ⓓ   19. Ⓐ Ⓑ Ⓒ Ⓓ   29. Ⓐ Ⓑ Ⓒ Ⓓ   39. Ⓐ Ⓑ Ⓒ Ⓓ

10. Ⓐ Ⓑ Ⓒ Ⓓ   20. Ⓐ Ⓑ Ⓒ Ⓓ   30. Ⓐ Ⓑ Ⓒ Ⓓ   40. Ⓐ Ⓑ Ⓒ Ⓓ

Practice Answer Sheet

Name _____

Date: _____ **Practice: #** _____

1. Ⓐ Ⓑ Ⓒ Ⓓ

2. Ⓐ Ⓑ Ⓒ Ⓓ

3. Ⓐ Ⓑ Ⓒ Ⓓ

4. Ⓐ Ⓑ Ⓒ Ⓓ

5. Ⓐ Ⓑ Ⓒ Ⓓ

6. Ⓐ Ⓑ Ⓒ Ⓓ

7. Ⓐ Ⓑ Ⓒ Ⓓ

8. Ⓐ Ⓑ Ⓒ Ⓓ

Date: _____ **Practice: #** _____

1. Ⓐ Ⓑ Ⓒ Ⓓ

2. Ⓐ Ⓑ Ⓒ Ⓓ

3. Ⓐ Ⓑ Ⓒ Ⓓ

4. Ⓐ Ⓑ Ⓒ Ⓓ

5. Ⓐ Ⓑ Ⓒ Ⓓ

6. Ⓐ Ⓑ Ⓒ Ⓓ

7. Ⓐ Ⓑ Ⓒ Ⓓ

8. Ⓐ Ⓑ Ⓒ Ⓓ

Date: _____ **Practice: #** _____

1. Ⓐ Ⓑ Ⓒ Ⓓ

2. Ⓐ Ⓑ Ⓒ Ⓓ

3. Ⓐ Ⓑ Ⓒ Ⓓ

4. Ⓐ Ⓑ Ⓒ Ⓓ

5. Ⓐ Ⓑ Ⓒ Ⓓ

6. Ⓐ Ⓑ Ⓒ Ⓓ

7. Ⓐ Ⓑ Ⓒ Ⓓ

8. Ⓐ Ⓑ Ⓒ Ⓓ

Date: _____ **Practice: #** _____

1. Ⓐ Ⓑ Ⓒ Ⓓ

2. Ⓐ Ⓑ Ⓒ Ⓓ

3. Ⓐ Ⓑ Ⓒ Ⓓ

4. Ⓐ Ⓑ Ⓒ Ⓓ

5. Ⓐ Ⓑ Ⓒ Ⓓ

6. Ⓐ Ⓑ Ⓒ Ⓓ

7. Ⓐ Ⓑ Ⓒ Ⓓ

8. Ⓐ Ⓑ Ⓒ Ⓓ